J

ANIMAL PSYCHOLOGY

Some other books by the same author

How Animals Talk

The Mind of the Dog

What Makes a Good Horse

The Conformation of the Dog

Veterinary Ophthalmology

Clinical Veterinary Surgery

The Examination of Animals for Soundness

Animal Vision

The Female of The Species

ANIMAL PSYCHOLOGY

A Book of Comparative Psychology Which Discusses the Behaviour of Animals and Man

By

R. H. SMYTHE, M.R.C.V.S.

Formerly, Examiner for the Diploma of Membership of The Royal College of Veterinary Surgeons

Illustrated by the Author

CHARLES C THOMAS • PUBLISHER

Springfield • Illinois • U.S.A.

CHARLES C THOMAS • PUBLISHER
BANNERSTONE HOUSE
301-327 East Lawrence Avenue, Springfield, Illinois, U.S.A.

Library of Congress Catalog Card Number: 61-13447

With THOMAS BOOKS careful attention is given to all details of manufacturing and design. It is the Publisher's desire to present books that are satisfactory as to their physical qualities and artistic possibilities and appropriate for their particular use. THOMAS BOOKS will be true to those laws of quality that assure a good name and good will.

Printed in the United States of America

CONTENTS

ANIMAL PSYCHOLOGY

Chapter I

LOOKING INTO THE ANIMAL MIND

As a result of the kind of education dealt out, nowadays, it is terribly easy to grow up with the conviction that humanity sits perched on a pinnacle, far above the heads of all the other animals.

Such a belief may prove entirely satisfying, until the time comes when emotion, fear or suffering, do their worst.

Inhibitions may then be thrown to the winds, and it may become only too obvious that the primitive animal lives not far beneath the human skin.

Animals vary so enormously among themselves, and when we speak of them in a general sense we may have in mind creatures as different as a water beetle and a chimpanzee. The behaviour of the chimpanzee may, at times, strangely resemble that of a human being, but, equally so, that of the water beetle, exercising its innate urges, may be not unlike that of the chimpanzee.

It behoves us to remember that the differences which exist between man and the so-called "lower animals," in their physical and psychological aspects, are purely superficial. Each one of them responds to stimuli, specific or otherwise, though the response of a human being may be entirely unlike the response of another animal to the same stimulus.

Accordingly, it remains difficult for a human observer to comprehend the behaviour of another animal so long as he expects its responses to run parallel with his own.

It would probably be equally difficult for the same observer, knowing full well what his own responses would be to any given stimulus, to forecast the responses of another human being.

The study of animal psychology differs from that of human psychology in one respect. Human beings are usually conscious of what they are doing, of the purpose of their actions, and of the results they hope, or expect, to attain.

Quite a lot of animal behaviour is carried out unconsciously, in automatic fasion. Its purpose may be unknown to the animal, which also may be quite unable to anticipate its possible result. One point of similarity, however, is that any form of behaviour

whether it be in man or another animal is "triggered off," to use a modern expression, by a stimulus. In animals, as well as in men, the termination of one piece of behaviour, may itself provide the stimulus which sets the next stage into operation.

The study of animal psychology was based for a long time upon human standards and expectations, with the result that a good deal of misconception arose, and it is only of late years it has been fully realised that animals are still primitive but none the less creatures capable of producing natural responses to stimuli. Human beings, on the other hand, live in a world of make-believe. Their responses, instead of being natural ones, like those of the lower animals, are conditioned by an artificial and unnatural way of life, and modified in accordance with the demands and requirements of what is still termed "civilisation."

It is no longer considered correct behaviour for a man to sneeze openly while in the company of his fellows, nor must he belch after a meal, or emit flatus in audible fashion. Other animals have no such inhibitions, nor would they understand why such natural behaviour would be greeted with human disapproval.

But we humans are ruled by what we term convention, although one may suspect that the majority of men may be far less conventional in their private lives. Behaviour, which would be considered impolite in social circles, might not be controlled so carefully if the individual found himself alone on a desert island. It is still frightfully easy, and altogether natural, for man to revert to the behaviour of the beast.

But even the lower animals may on occasion exhibit many of the characteristics of behaviour observable in members of the human race, and, individually, animals may display temperamental variations very like those we encounter among our own friends and acquaintances.

Animals love, fear and hate, in the same way as we do. They trust and mistrust, exhibit staunch loyalty as well as indifference. They become attached to their homes, and develop nostalgic symptoms when removed to strange surroundings. A great deal concerning the emotions of animals can be learned within the walls of a boarding kennel for dogs and cats, during the holiday season. If we are correct in the interpretation of animal expression we may observe indications of joy, grief and anger. We can determine whether animals are happy and confident, melancholy or riotous. We can pick out the natural optimists and the born pessimists as easily as we distinguish the dainty animal from the greedy one. We must base our observations upon what is obvious and factual and refrain always from guessing what goes on in the animal mind.

Those who deny the ability of the trained observer to recognise the emotions and feelings of an animal with which he is familiar, should ask themselves if they are quite sure that their *own* powers of observation are well-developed and that their devotion to psychology is not interfering with their understanding of animals.

I feel confident that in claiming ability to recognise, in a great many cases, the mental or emotional reaction of an animal by a study of its expression, its posture and general behaviour, an experienced observer is not necessarily guilty of anthropomorphism, provided always that he is never swayed by his *own* emotions, and is putting into practice the knowledge he has acquired by long experience of, and close contact with, many kinds of animal.

At the end of Chapter I, I append a short summary of the emotions in animals and how they may be recognised.

Studying Animal Behaviour

If you intend to study the behaviour of animals in a truly scientific manner you will have to start at the beginning with simple examples and minor problems. You do not need a zoological collection, nor do you require a lot of intricate apparatus. A dog and a long lead, a canary in a cage, or a tame white rat, any one - or all of these if you are ambitious - will provide the essentials.

When you have ascertained, without any doubt, what these animals do, why they do it, what motives or purpose lie behind their actions, and what sets each form of activity into operation, you will have accomplished a very great deal.

On one unhappy occasion in my more youthful days, I wrote a series of articles for a scientific magazine, in which I described the everyday behaviour of an Airedale I owned at the time, with my own analysis of such behaviour based on the principles of research outlined in the above paragraph. The articles came back, with the comment that what I had related described ordinary behaviour of no interest whatever. What was required was an account of *extraordinary* behaviour, the kind of story calculated to interest and *stimulate* readers.

Once again may I point out that the kind of behaviour about which we know the least, which we should study far more closely at first, is the ordinary variety which we can observe at almost any time. Because it is so plain for all to see, and so familiar, it often goes unnoticed.

It is always so easy to overlook the obvious in the search for the obscure.

I think it is more satisfactory to observe the behaviour of an animal first without too close a study of scientific volumes, and from one's own observation develop some theories, rather than read the generally accepted theories and then try to persuade oneself that they credibly explain all one observes. It is a good practice to keep a notebook, the pages of which are divided into three columns. In the first describe what you have actually observed - nothing more or less. In the second column enter your conclusions as to the purpose of the behaviour, if any; what you think acted as the stimulus, and the result of the behaviour. In the third column make any suggestions, followed by a query mark, in order that you may again consider these when you have acquired further experience.

If I adopt the former plan and advise placing greater reliance on personal observations than upon established theories; if sometimes I place the hypothetical cart before the rather too positive horse, I ask forgiveness for so rebelliously departing from the standard formula. I shall confine myself to the study of a very few animals, most of which are well-known to us all, rather than deal with a wide, zoological collection. I will endeavour to show how the average animal responds to what are termed 'specific stimuli', as well as what it does on occasions when it is confronted with the unexpected.

Living with Animals

The only way to study animal behaviour, as it really occurs, is to *live* with the animals, preferably in their natural environment, but as this is not always pleasing to the animals, the sole alternative is to watch very closely the behaviour of any kind of creature that can be persuaded to live with oneself. I can claim, probably, that I have taken into my home, as members of the family, as great a variety of strange animals as most people have done, and I have found that when they have overcome their initial shyness and gained confidence, they accept one's dwelling as their own, display all their natural characteristics, and provide very entertaining company.

As a boy, I frequently kept fourteen or fifteen varieties of animals at one time, and as I was not allowed to keep them indoors, they all shared a large loose box in our stables. It was through them that I made my first discovery, one I have always regarded as being important. It was that animals which in their wild state would have been mortal enemies, such as a ferret and a rabbit (on one occasion a polecat and a rabbit), when they were reared together from early infancy, away from their mothers and maternal tuition, would grow up together in friendship and live in perfect harmony, with no desire to kill, and no fear of being attacked.

In this way I collected soon after their birth, a fox, a badger, a ferret, several hedgehogs, a number of white and coloured rats, a smooth fox terrier, two green monkeys, a donkey and a cat. As I have already made mention of this assorted family in *How Animals Talk,* (Country Life Co.), I will say no more about them, other than to mention that they lived and slept together, and all of them, having made my acquaintance very early in life, accepted me in place of a father.

Very few animals come into the world with the inherent urge to kill, in spite of the fact that when carnivorous they accept and eat raw meat if it is offered to them. A number of wild carnivora accept or adopt an alternative diet when circumstances make it necessary. Both dogs and foxes will live in captivity on a diet of barley meal cooked with milk and vegetables, while wild foxes will rob orchards and fruit gardens, and eat blackberries, nuts, raw turnips and potatoes, with apparent relish.

I know one orchard in Devonshire in which, during the summer and autumn, badgers climb the trees almost every night, going up into the topmost branches to steal the apples.

In their normal surroundings fox cubs are *taught* to kill their food. The father fox brings in the game alive, or at least half-alive, and the mother teaches the cub to hold the rooster or pullet by the front portion of its body to prevent it struggling, while it gnaws away at the hinder end with its teeth. Neither the vixen, nor the cub, has the faintest knowledge that such treatment causes pain, nor would either be capable of mentally translating itself into the role of the victim. Such a chain of connected thinking would be far beyond its mental ability. Similarily, very few kittens, taken from their mother at an early age and reared in the household circle away from all other cats, will catch and kill mice, although they may roam and play with them without any attempt to bite them. One of my female kittens, on achieving maturity, actually adopted a family of black and white rats, and suckled them on real milk, although the kitten had never at any time mated.

In the same way, two greyhounds I reared by hand from a bitch which died a few days after whelping, developed into fine dogs and won many prizes in the show ring. Taken out to see how they shaped at coursing when they were fifteen months old it was noted that they ran with the hare and even made attempts to play with it, but none to bite it, or impede its progress in any manner.

My buck rabbit and the ferret held each other in the greatest respect, but it was the ferret who took cover between the fox and the badger whenever the buck stamped his feet in anger.

Apart from this tendency to overcome natural enmity between varying species, domestication does not lessen the mentality of animals - on the other hand, association with kindly human beings develops intelligence in them. Animals learn quickly from human foster parents in their new homes. But it must be remembered that anything of the nature of what Lorenz termed "imprinting" occurs only when the very young baby animal is adopted by the human being on whom it first sets eyes, It may then accept that person as its parent figure.

Recently I heard of one puppy which was adopted (when its mother died at birth) by a lady who wore a large blue overall. For several months the puppy would show marked friendliness to its protector whenever she wore the overall, but would ignore her completely if she appeared without it. It soon became clear that the "imprinting" concerned the overall rather than its wearer, and this was confirmed when an entire stranger of the male gender donned the overall, and after playing with the puppy, carried it home with him. Unfortunately, he left the overall behind, and the puppy refused to have anything to do with him in its new environment. However, he returned to the puppy's birthplace and secured the overall and from that time all went well.

Young animals such as calves, lambs and goatlings, sometimes imagine, apparently, that they are of the same species as their parent figures, and when the breeding season comes round will expect them to carry out a mating programme. One cockerel reared in this way used to strut around its mistress with the wing-kicking behaviour typical of the amorous rooster, obviously inviting her to step into the nest box and lay eggs.

A young ram lamb finding his mistress unresponsive, butted her as she was looking into a barrel of pig-swill. Into this she fell head first, and was only rescued in time by the fortuitous arrival of one of the farm hands.

Although wild animals may often be conditioned into becoming household pets, they are apt to forget their acquired manners on very small provocation, and revert to their natural behaviour. This is particularly the case during their normal breeding season, when the urge to escape to a highly congenial environment becomes more pressing than their love of comfort and human companionship. If thwarted in this enterprise they show frustration behaviour, which, put plainly, means that they bite their owners, or protectors. The lad who has carried a grass-snake in his jacket pocket and fondled it for weeks by stroking it with hand in pocket, is surprised one day when it grabs his finger in its mouth and refuses to let go. He does not realize that the snake is feeling the mating urge and has now no further use for pockets.

Domestication widens the scope of the animal's experience and introduces it to situations and problems, which it would never encounter in its wild state. It has no instinctive answer to many of these and it is by watching how it tackles the problems that one learns a great deal from unusual pets.

There is, however, the human angle to consider. Very few people are so interested in animals and their behaviour that they are willing to sacrifice their leisure and pleasure in order to watch them. In addition, many adopted animals, especially monkeys and the wild carnivora, become difficult to manage, or positively dangerous, when they attain puberty. Those which remain friendly with their parent figures, may be otherwise than friendly with callers.

I must, however, repeat the statement that when and while it can be carried out, animals can be studied far better in domestication than when they are kept in laboratory cages.

Unfortunately, the scientific world contains a great variety of cages. Those intended for animals are made of metal and are highly hygienic. They possess no water or gas taps that monkey fingers can turn on, but they do possess tiny pipettes, from which the prisoner can lick droplets of water when thirsty.

The inmates are usually scientifically inbred, mentally-defective animals, unloved and unwanted. They exist in an unaccustomed environment, live on a carefully devised but uninteresting menu, and are known not by names but by numbers.

So far as natural, uninhibited behaviour is concerned, these frustrated, unhappy creatures can provide us with little knowledge. They have sufficient intelligence to turn left each time they receive an electric shock on the right side of the body - but an earthworm, which is almost devoid of intelligence, does precisely the same thing.

If young students of animal behaviour would endeavor to interest themselves in real animals, not in units in a prison; if they could be persuaded to adopt young animals, give them affection and be loved in return; rear them as they would children in their own homes, their conclusions regarding animal behaviour would be far more valuable, and their understanding of animals so much more intimate. The greatest difficulty the student may experience may be in arousing an enthusiasm, equal to his own, among the remaining members of his family!

The Scientific Approach

Many of the earlier psychologists, but by no means all, favoured the *mechanistic* attitude towards the study of animal behaviour, paying little or no attention to the animal's psychological constitution, regarding it as a piece of machinery requiring housing and stoking, and little else. The younger generation nowadays tend to favour the *naturalistic* approach, with excellent results. The fact that so much is being done in the photographic field indicates that animals can be approached and viewed in their natural habitats if suitable precautions are taken. Scientific observation can now be combined with healthy, outdoor exercise - a fact which is gaining many recruits.

Purposive Behaviour

When we have learned sufficiently about them to accept the fact that animals in no way resemble machines, and that they are imbued with personality and a marked degree of sensitivity, we may commence to determine how much of their behaviour is *purposive,* that is to say carried out with a definite aim, and how much is conditioned by other circumstances. We shall also discover, probably, that some part of it is *automatic.* We will discuss this type of behaviour in the next chapter in its connection with what, for the moment, we will term *instinctive behaviour.*

We shall learn as we read further that the behaviour of all kinds of animals, not excluding that of ourselves, is influenced, or conditioned by, the secretion of chemical substances termed "hormones," into the blood stream. These substances are derived from the endocrine - the so-called "ductless glands." The glands which produce the most obvious seasonal effects are the sex glands - the testes and the ovaries - which control the reproductive cycle, and all the phenomena leading up to mating.

We may experience some slight difficulty in determining where purposive behaviour ends and at what point behaviour is taken over and directed by automatism, based on innate impulses, or instincts. We are able so much better, when we live with the animal, to determine exactly what the animal is doing, what it aims to do, why it should be impelled to act in the way it does, and what it finally accomplishes. We must try to determine how much of its behaviour is instinctive, and therefore in some degree automatic, and, also, to what extent the animal is conscious of what it is doing. How much does it see, and hear, and feel, and what is its reaction to the sum total of its sensual perceptions? We must also guard against falling into the mistake of imagining that because an object, or something which happens, conveys an impression to one's own mind, it will necessarily convey a similar one to the mind of an animal. It may, in fact, transmit an entirely different one.

Life from Another Angle

Our respective values may also be quite dissimilar. *We* should not welcome the arrival of a pad of cowdung on our best carpet, we might exhibit disgust. But to a retriever puppy, especially when hungry, the pad might convey the impression of food. That it might contain useful minerals and vitamins would be beyond its understanding, but its instinctive impulse would be to sample it and find out if it were edible or not. The same puppy might retreat from a ripe and dainty portion of Camembert, which many an epicure might enjoy. We might take delight in an old masterpiece, framed and standing in our hall awaiting hanging. Our dog would see nothing but a wooden frame encircling a patch of greyness, far less interesting than the nearest lamp-post, a useful article his master from some stupid oversight has never yet contemplated erecting in the hall. The dog is practical by nature and completely devoid of all aesthetic taste. Beauty is for others, the dog prefers beef. It has a far higher food value.

I am sure it would help you a great deal in your study of the world a dog or a mouse lives in, if you could convert yourself into the particular animal you are considering, if only for a few moments. You would then be in a position to compare its size and shape with your own, and study its posture, balanced on four feet instead of upon only two. This would enable you better to appreciate the general circumstances connected with its life, and how the animal views its own immediate environment.

Would it be too much to ask you to imagine that you have changed overnight from a well-developed human being into a tiny mouse? Until today, in your human shape, you have become accustomed to surveying a wide area of land and gazing at a far distant horizon with your eye level at a little over five feet above that of the ground. In your human shape your eyes observed a great deal of detail. You were aware of the nature of most of the objects you perceived and the purpose of their presence. You were able to focus your gaze on to those which were close, and you could see other objects quite clearly, at a considerable distance from your eyes, anything up to half a mile let us suppose. You had become quite used to seeing other animals, large or small, human or otherwise, moving about in your vicinity, and you could decide at a glance whether they appeared friendly, indifferent to your presence, or hostile. Being an intelligent human being you would probably have retained adequate means at hand to defend yourself in case you were attacked.

In your new role as a mouse, you will become aware, at once, that your eyes are now an inch only above ground level, and that the tufts of grass you used to kick with your toes, have suddenly taken on the appearance of small plantations, obstructing your vision almost completely. Your visual acuity has also deteriorated in an alarming manner, with the result that whenever you do manage to obtain a clear view around you, the horizon is represented by a circle with a radius of only six feet from the tip of your nose.

Apart from a few ground enemies - the stoat and the weasel, the cat and the dog, and perhaps occasionally an adder - most of your danger comes from overhead. The owl or the hawk may drop down upon you before you are aware of their presence.

Reflexes and Responses

Only a short while ago you were a tall human being. You had every opportunity to look around you and sum up the situation. You could reason in your mind whether it would be safe for you to cross the field ahead of you, seeing that, very obviously, it contained the farmer's bull. Your intelligence would advise you to arm yourself with a stout stick - and off you would go. But, as soon as you had entered the field and closed the gate, the bull might let out a loud bellow and come careering towards you. Would you stand your ground, and reason what would be the wisest course of behaviour to adopt? Not very likely! Your inherent impulses would take charge. The muscles which operate your limbs would contract in ordered sequence. You would not actually decide that they should do this because before you could map out a course of action your reflexes, set in operation by a visible stimulus (the bull), would have taken charge of the situation, and then, to your great relief, you would find that once again you were on the right side of the gate.

During the short period of time this would take you would have ceased to be a reasoning human being and you would have apopted exactly the same type of behaviour any other animal would have practised. In fact, you would have done precisely what the mouse would have done if it had come face to face with the cat - bolted!

Perceptivity

The mouse, in its natural environment sees very little, but it hears a lot. In addition - and this is very important- like all animals that live close to the ground, with two fore feet linked up to two hind feet by the spinal column, the mouse is very sensitive to earth vibrations, even those which originate a considerable distance away.

It is probable that a horse, a zebra, or any of the ungulates which possess solid, horny hoofs, can sense the vibrations caused by the farm hand's hob-nailed boots walking along a hard road surface, when he is still a quarter of a mile distant.

It is likely that carnivorous animals, rodents and lagomorphs (hares and rabbits) can "feel" vibrations through the earth at greater distances than the hoofed ungulates, although a great deal depends upon the degree of humidity of the soil. In cold, frosty weather rabbits and birds, too, can detect footsteps upon turf, which give but little audible warning, at very great distances. The thrush and the blackbird have been credited with such acute hearing that either can detect the movements of a worm underground at a distance of several feet. It is quite as likely that the bird "hears" ground vibrations, set up by the worm, transmitted through its feet and limbs. The bird requires to be *on the ground* in order to detect the presence of the worm emerging from its burrow. Nobody has reported that thrushes "hear" worms whilst they are perching in a low bush. More often it is a blackbird which sits in the bush, watching the thrush capturing the worm, or smashing the snail shell upon an anvil of stone, ready to dart down and snatch the booty from the thrush without checking the speed of its flight.

The mouse, short sighted and defenceless as it may be, is still very capable of self-preservation. Like many other small animals not very high in the scale of intelligence, the mouse is nevertheless highly sensitive, and its responses are extraordinarily rapid. When danger threatens, the mouse does not stop to think. In this way it behaves much as a child would do when it flees from danger. But the wild rush of the child may lead it into greater danger. There are no motor cars in the world of the mouse and so it reaches its hole quite safely - which is why there are still so many mice!

If one goes to the other extreme and considers an animal such as the giraffe - a very tall ceature which enjoys an extensive view of the countryside, one may regard it as having no need to beat a precipitate retreat from any other animal, as the mouse finds it expedient to do. Not only is the giraffe able to detect any possible enemy but being fleet of foot and provided with long, powerful hind limbs, it can deliver a kick at a pursuer which even a lion holds in respect.

Intelligent Response

Responsive behaviour varies with individual characteristics. As we have just observed, human beings, in spite of their highly developed brains, are still physically vulnerable, and depend for survival upon rapid, unconsidered responses to specific stimuli just as frequently as they rely upon their ability to reason. In this respect they remain on a par with the animals beneath them in the social grade.

But man in a civilised country expects to receive a certain amount of protection, provided by his fellow men against common enemies and dangers, and accordingly he is apt to become careless.

But the mouse, safely back in its hole, need worry no longer. It has no occasion to meditate over the vagaries of behaviour observable in other creatures, or upon the idiocy or intelligence of the human species, nor does it indulge in any analytical post-mortem regarding the circumstances which enabled it to escape the cat. There would be no advantage to the mouse if it possessed a higher degree of intelligence; it is not obliged to think in order to live. In the mouse world brains carry no sales potential and in any case the mental ability of the mouse falls short of imagination, or conscious thought. Silent meditation and deep thinking are the prerogative of humanity. Look around the world today and observe what our intellect has achieved.

But the mouse is a lowly little creature and one may accord to many animals a somewhat greater degree of mental acuity. Its intelligence cannot compare, for example, with that of the dog, or of other animals which have lived in domesticity, shared human companionship, and acquired at least a small degree of human wisdom.

But there are other animals living in the world of today which have had ample opportunity to cultivate thought processes if it were possible for them to do this without human aid. The lion, the elephant and the crocodile have no natural enemies other than man, nor do they ever stand in need of a meal. With unlimited time at their disposal, not one of these animals has developed a standard of intelligence while in the wild state, greatly exceeding that of the much handicapped mouse.

But the elephant, owning one of the smallest of brains as compared with its body weight, when once it has made close association with mankind develops a most extraordinary degree of intelligence. It works for its human master and even thinks out ways (it would seem) of accomplishing a tricky job of work more cleverly than the man who normally directs its operations. This is another indication that animals seldom lack some intelligence, but are deprived of education in the wild state, and are quite capable of learning from animals more intelligent and more experienced than themselves.

In this way there again appears to be some similarity between the minds of animals and those of children.

It is easy to establish mental contact with the majority of dogs - and physical contact with a horse - and from the "feel" conveyed from hands to mouth through the reins, and from body to body through pressure of limbs on saddle, to make some sort of contact with the horse's mind.

But usually one can make only a one-way contact with any member of the feline race. Cats, large or small, are subservient neither to man or beast, and although they may tolerate human companionship with an eye to comfort and sustenance, they obey no commands other than those which conform strictly with their own desires.

At all times and in all circumstances the domestic cat retains complete independence, just as in the jungle the cats, large and small pay homage to no other animal. In nature dog eats dog, but who ever heard of lion eating lion? Cats live in a world of their own in which they are sacrosant.

One can exact implicit obedience from an Indian elephant when once it has submitted to the domination of man, but the camel, rather like its fellow-slave, the ass, is always a rebel at heart. There are many animals, apart from the cat, contented only while they are getting their own way, and these are to be found in every species, high or low in the biological scale.

The Influence of Environment

We cannot see far into the animal mind unless we understand its environmental state. When you, yourself, visualize your own surroundings you picture your home, your office perhaps, maybe your consulting room, and in all probability your favourite golf course.

Man is fortunate in possessing a greater ability to choose, and to select from a far wider choice of environments than other animals, lower in the social scale, can do. He has far greater opportunity to plan his own environment rather than accept unconditionally whatever Nature may be pleased to provide. In addition, he can select his clothing and vary it in accordance with the state of the weather, whereas the animal owns only one suit, and at certain times of the year this becomes very threadbare.

Man may decide to live in an air-conditioned building at one prescribed temperature throughout winter and summer, and if he is a lover of fresh air and open spaces he can choose the plains or valleys, or he can move to the mountains, or to arctic conditions if he prefers them. He can build or he can demolish —his chosen environment is almost invariably an artificial one.

Animals are sometimes able to move from one environment to another, or they may be compelled to remain in one place and make the best of what exists. That animals do often continue to thrive and multiply amid what to our minds would appear highly unsatisfactory conditions, is an indication that they are the survivors of a large family. Those which could not exist in such surroundings either died or failed to reproduce their kind.

When animals have a choice, they select the environment which will fit in best with their natural manner of living, one which will enable them to rear their families, and increase their numbers.

Man aims at comfort—animals are never fussy about the amenities, so long as they can live in safety.

Every animal from mosquito to man values environment differently, according to how it fits with certain special needs.

Environment is a term which embraces a great many kinds of bodily surroundings, and we employ the term "territory" to describe that portion of its environment which the animal regards as its home, or headquarters; over which it holds sovereign rights.

Territory—which we shall discuss more fully in a later chapter—may be a masculine, or less frequently a purely feminine, possession; or it may be shared by a male and a female in marital association.

As previously mentioned, the ability of animals to move from place to place varies enormously, and although even limpets are far from stationary in their habits it is probable that an average dog covers more ground in an hour than a limpet does in a lifetime; yet a swallow may move from one continent to another overnight.

Later, too, we shall have a good deal to say about bird migration. At the moment I would like to consider matters much nearer home, and describe in a very simple way how different is the impression the same area of land, or other type of locality, may convey to a man and to another animal. At the risk of failing to exhibit cleverness by discussing more remarkable examples of animal behaviour, I will describe the simple reactions of a dog, in the hope that these will help us to understand canine behaviour.

I would like you to visualise my own Scottish terrier taking me for a walk at 7:30 a.m., regardless of the weather. I put it this way because we set out from home, with 'Whisky' straining

at the lead and towing me along the pavement, while he arouses the neighbours with his joyous barking; a clear indication that if one wishes to obtain real pleasure from living it is not essential to be too intelligent.

Our first port of call is always a piece of waste ground, a hundred and fifty yards from our own front door.

To myself and my friends, this area of land, half-an-acre in extent, constitutes an eyesore—a blot on a respectable residential neighbourhood. Among ourselves we visualize it not as a canine meeting-place but as a potential site for the erection of a house, a bungalow, or a dozen lock-up garages. You will note that our thoughts are centred around money, a very necessary human commodity, but unvalued by the dog. Whisky would regard the site through quite a different pair of spectacles. The mind of a dog is concerned solely with "doggy" business.

To every different type of animal the business of living lends importance to certain features of its environment which overshadow all others. They are aspects peculiar to the animal's needs and to us they might seem unimportant.

An environment which would seem ideal to a family of otters might hold no charm for a pair of badgers, and the stubble field which suits a covey of partridges might be equally unpopular with a gaggle of geese.

In our own lives we may decide whether we would prefer to live in Mayfair or Limehouse, according to our needs. As the human race progressed it chose flats in skyscrapers in preference to damp caves, although it is probable the cave-dwellers grumbled less about housing accommodations than modern folk do.

The fact that the plot of ground we have been discussing is insanitary, a dumping ground for various undesirable objects, and the home of at least one pair of rats, only enhances its amenities from the canine viewpoint. If Whisky were capable of contributing his impressions concerning this plot of ground, he would undoubtedly lay emphasis upon certain features which would be carefully excluded from the advertisements of every estate agent interested in its sale. I am sure that every dog who frequents the spot would give full marks to the multitude of entrancing odours it is capable of providing—many of which have, as yet, never been completely analysed by any of its canine visitors. All of these perfumes have yet to be catalogued and their particulars filed away in scores of doggy brains for future reference.

In writing of odours I used the adjective 'entrancing' quite unashamedly. I am not at all afraid of those exacting people who will accuse me of ascribing to dogs a near-aesthetic sense, which enables them to revel in their olfactory perceptivity. True appreciation of art is not confined to vision; a blind man may find delight in handling a choice piece of pottery, or in letting his fingers glide over a massive block of sculpture.

No man could obtain the same degree of satisfaction from a wayside odour as Whisky does, partly because the human olfactory sense is almost completely undeveloped when compared with that of the dog or even with that of other quadrupeds, and fishes.

When Whisky buries his nose in a tuft of faded yellow grass, when his whole body stiffens and he shakes from head to tail with excitement, and when he remains rooted to the spot heedless of man or beast, he is not merely inhaling a scent. He is splitting it up into its fractional parts and estimating to the nth degree the exact physiological milepost attained by the interesting creature who has left so delicate a bouquet in her wake.

And if one must really know the origins of the delectable perfumes which more commonly entrance dogs in this fashion, the answer is a simple one. They are: (a) Urine, (b) Vermin.

In the case of (a) the subject of the entrancement will usually be male, and the origin of the perfume, female. In the minority of instances the reverse holds good. When the female is becoming exciting in the amatory sense, the male may commence also to exhibit interest in urine of the male order also. Close concentration upon this will then provide a complete list of the male suitors who constitute the immediate opposition.

The word "vermin" in the opinion of the dog embraces a mixed bag of rats, mice, voles and hedgehogs, the kind of small game it *might* be possible to find within the space of half-an-acre of derelict land, if one were *very lucky, indeed.*

Lying behind this pressing urge of the domestic dog to seek out and kill, or worry, less fortunate creatures is the *innate need* to seek food; and although Whisky would turn up his nose at any creature in the above list—after he had shaken all life out of its body—he would undoubtedly experience a warm glow of satisfaction as the result of setting into operation a programme of instinctive 'doggy' behaviour, and having carried it through to its bitter end.

The fulfilment of any *need* is always associated with pleasure, although the pleasure may be preceded by pain in some instances. As we shall learn a little later the performance of an *instinctive action* makes little impression on the animal mind—in fact the animal is more often unconscious that it is carrying it out, and why!

Pathways to Perception

By watching the foregoing performance we shall discover that at least one animal, a dog in this instance, makes use of his sense of smell rather than his visual sense for the fulfilment of two innate needs:—

(a) The search for a mate in a physiological condition ready, or almost ready, for mating; an operation necessary to the perpetuation of the species, even if the two participants are unaware of its purpose.

(b) The provision of bodily sustenance.

So fas as (a) is concerned, the "ripeness" of the scent may convey to the canine mind the knowledge that some little time may pass before the mating urge will achieve consummation; but recognition of the existence of the perfume, and of its whereabouts, will create a desire to revisit the spot at intervals, as frequent as may be possible. A creditable persistence may not only lead to a better understanding of the progress of the charmer's physiological condition but it may by some odd chance bring about a meeting with the charmer herself, an introduction which is sure to provide material for a story with the usual happy ending.

If Whisky were some animal other than a dog, I might be able to make a few remarks concerning (b), the provision of bodily sustenance for a mate and a prospective litter of puppies, also aided by the dog's olfactory sense.

Unfortunately, however, the male dog is a philanderer, and after his lovemaking is done, he goes upon his way, filled with a sweet indifference.

According to the mechanistic school of animal psychology emotion plays little part in the make-up of an animal. This takes me back to my use of the word "entrancing" when I described a dog's reaction to an extremely attractive perfume concealed within a tuft of grass. I think the use of the adjective was justified. It is said that in the blissful state of entrancement, we humans merge into a dreamy condition, in which we are filled with rapturous delight.

I feel confident that any unbiased observer who cares to join me in my early morning walk and observe Whisky's complete absorption in his latest exciting discovery will agree with me that we can interpret the objective signs and symptoms he exhibits in no other way.

Uexkull in *The World the Dog Lives In,* demonstrated that every species of animal had its own particular sphere of perception and influence, provided by its own organization. No being of a different type of organization could ever make use of another's sphere.

He showed that a tick, blind as it is, sees nothing of the world around it, and asks for nothing better than to live in a narrow belt of warmth and a wide belt of scent.

Like the dog the tick has one main absorption—odour. It is interested principally in the butyric acid secreted in the sweat of the animal which acts as its host.

In a somewhat similar way the mosquito remains suspended in an atmosphere which holds for it no great charm, but it is strongly attracted to a nearby animal, usually a human being, not through the medium of a perfume, but by the tiny degree of warmth radiating from the human body over an intervening space of several feet.

This ability to recognise so minute a fractional rise in air temperature, amounting only to perhaps a hundredth part of a degree, at close range, is a characteristic the mosquito shares with a number of other unpopular creatures, which include the bed bug, the flea and certain snakes, the pit vipers. These few examples may be sufficient to show that all animals are not dependent upon the same sense receptors and that a pattern of animal behaviour may be set into operation, or triggered (to use the modern expression) by a variety of factors.

The Matter of Space

Whatever may be the advantages of domestication there is no doubt that the simple life in natural surroundings, with perfect freedom, provides the kind of existence which the majority of species would select. Domestication is possible in a great many varieties of animal, but it is an unnatural way of living and it introduces the kind of problems that only a few species seem able to deal with successfully. The greatest problem that confronts any domesticated animal is always its owner.

The animal which lives best in a state of domestication is one which, like man, is normally a member of a community, in other words of a pack, or herd, or flock, as the case may be. Such a one accepts one person as the pack leader, or "boss," and by instinct obeys his or her commands without any attempt at deciding whether they are reasonable or otherwise.

But all species of animals which live in freedom are not fond of the world with its wide open spaces. There are many which are endowed with an agarophobia, a morbid dread of the open. They fear the fields and their freedom, and desire nothing better than to creep into a hole in the ground, or into any place where they can feel its walls in contact with their body surface.

The worms, which inhabit closely fitting holes in the ground, the moles, rats and mice, are examples. We encounter other species, though, which inhabit burrows but are also not averse to a little excursion into the outer world. Such are the foxes, the badgers and the rabbits, but all of these do most of their prowling by night.

It is usually the smaller animals which exhibit true agarophobia, and these are the ones which are hunted, while the hunters, especially those found in the cat family, are either night prowlers, or members of the wolf and dog species, which are arythmic animals, able to cope with daylight or to hunt successfully in a dim light, such as morning or evening twilight.

It is quite obvious that the instinct to hide in holes is linked with its survival value for not only does the agarophobic animal dislike the open country but it has good reason to do so, for in the open it is an easy prey, and any one of these animals which overcame its dread of space would soon cease to exist and to produce more of its kind.

One must not surmise, however, that any animal is conscious of the fact that it may encounter destruction. It is ruled simply by its innate agarophobic impulse and, as I will explain in a later chapter, death is something of which an animal has no knowledge.

We ourselves possess some traces of this primitive agarophobia and we show it in our love of clothing, of being well swaddled in bedclothes, and there are records of many such persons ranging from hermits and cavedwellers down to Diogenes, who discovered happiness in a tub. But although we favour corsets, belts, and tightly fitting garments—a survival of our primitive urge to "get inside" some form of shelter—at least some of us have also developed a claustrophobia, a fear of confined spaces. It is likely that this is not so much a primitive instinct as a "conditioning" of the mentality, which now harbours a fear that the physical body may be buried alive, a theory which explains why a person confined in an enclosed space feels that the walls are growing closer to one another and that the supply of air or oxygen is rapidly diminishing. The larger apes, the gorillas especially, do not appear to have developed this claustrophobic tendency. The male gorilla invariably sleeps back to the wall and has no fear of entering a small space or living within it. Being able to feel a solid rock on three sides of the body, one of which is one's back, is one method of avoiding being taken by surprise.

Habit

When we look a little further into the minds of those animals dependent for the continuation of their existence upon their own wariness we find that their security measures are based largely on what we may term "habit." When we are asked what we mean by this we can only reply that the term describes a regular form of behaviour which has developed into an ordinary course of conduct. In other words a pattern of behaviour which has so far operated successfully in establishing its survival value has now become automatic, rather than purposive, in its nature. Such behaviour has become conditioned by repetition.

We ourselves often declare that "we are getting into a rut," and this expression is equally applicable to a great many wild animals living natural lives.

Thus, a rabbit or a hare which has become used to travelling through a well-used run, traversing a cover of ferns or brambles, will persist in following the same pathway for as long as it continues to provide protection and concealment. The animal feels its skin constantly coming into contact with the walls of the run, even if these are only bracken or bushes, and it derives from this a feeling of security. This may eventually prove a false one since poachers and trappers, as well as other kinds of creatures, may take advantage of the animal's habit.

I knew one young farmer in Corwall who supplied live badgers to laboratories. He caught seven in one week by hanging a running loop of chain, after the style of a rabbit snare, in the end of a run used by a family of badgers, just at the point where they jumped down from a low wall to the ground below.

Although the loop was quite visible, each badger was so absorbed in following the exact track that it entirely ignored the chain and all seven contrived to place their heads through the loop before they jumped down from the wall.

A rather curious instance of a pattern of behaviour developing into a habit was recently exhibited by a terrier owned by one of my friends. The owner went out to work each day leaving his dog in the care of his wife. Unfortunately she was obliged to go into hospital, and in order that the dog should not be left alone I offered to look after it by day and it was arranged that it should be fetched by its owner each evening. For a fortnight the dog arrived at my house at 8 a.m. each morning in charge of its owner, who was then on his way to work. The dog would come upstairs to my sitting room and be regaled with sweets and biscuits before I adjourned to my study to read and write. On the evening of the fifteenth day the owner did not arrive to collect his dog, and so he slept that night in my sitting room. On the following morning he was obviously ill at ease. I made no effort to go through the usual routine of providing sweets and biscuits because I wanted to see how the dog would respond to the new situation. I had not long to wait. After a little while he went downstairs and found that the front door was open. He went into the garden, then trotted a few yards down the pavement, turned, and barking joyously galloped his way upstairs to my sitting room, sat up on his stern in begging fashion and awaited his sweets and biscuits. Having received these all was again well. His behaviour pattern was restored.

I do not suggest that the dog went downstairs and out of doors entirely in order that he might stage his entry in his usual way. Probably he had other motives. But the fact remains that once having regained the pavement this acted as a stimulus and the dog's usual pattern of behaviour came into action as the response.

I should not have enjoyed this experience if I were concerned only with laboratory animals kept in cages.

One other rather striking example of the formation of a habit was provided by a horse working in a brewery wagon.

This animal cut its foreleg just below the knee and in the course of treatment the lower part of the limb was encased in a white bandage. When the horse returned to work the bandage was worn regularly for a few weeks to prevent any breakdown of the wound which had now healed. On the day the bandage was finally removed the horse refused to draw its load away from the brewery yard, until someone suggested that he might be missing his bandage. When this was wound around the limb again the horse started off as usual and completed the day's work without a hitch.

He continued to work, wearing a bandage, for a number of years, but on each occasion an attempt was made to discard it the same thing happened. The horse would refuse to pull the cart an inch until the decoration—for by now it could be nothing else—was restored.

MANIFESTATION OF THE EMOTIONS

An *emotion* is a state of mind associated with excitement or perturbation. It is expressed by strong feeling, which in the higher animals, and in a great many of the more intelligent animals below man, produces some external manifestation.

The stirring of an emotion creates an impulse, and this may immediately set in train a pattern of behaviour which expresses the animals response to the type of mental stimulation associated with the particular emotion.

If we hope to see into the mind of any animal we must endeavour to appreciate its reactions, both to its environment and towards other animals of the same or different species. We do this by watching its general behaviour, but more specially by interpreting from its manifestations the kind of emotion it is experiencing. Before we can do this we must possess knowledge and experience of the signs by which animals express the different form of emotion, so that we will know whether they represent joy, sadness, anger, bragadaccio, aggression or fear. Among animals we seldom encounter any indications of self-pity, nor do we find any hypochondriacs. On the whole, animals are extroverts and materialists, entirely devoid of sentimentality in the lower orders, although when we examine the more intelligent animals, the monkeys and the domestic dogs, we do occasionally find individuals which appear to border on the sentimental. But the true scientists refuse to accept what they term "such nonsense." Two baby monkeys cuddled closely together are merely keeping warm, (even when the temperature outside is 98°F.), and two lovebirds canoodling in a cage are exhibiting hormonal urges. It would be easy to provide similar explanations for all human conduct and deny that such a thing as sentiment exists.

Animals, nevertheless, are outstandingly practical in their attitude towards living, and moreover they are outstandingly truthful. Whether they are always equally honest according to our standards, cannot be decided until we are better acquainted with their own.

As the various animals display the state of their emotions in different ways, I will describe very briefly a few of these as shown by some well-known animals. It must be borne in mind always that animals learn far more of our moods and intentions from observation of our expressions than we are able to learn from a study of the animal. They far exceed man in sensual perceptivity and seem to know exactly what is going on in the human mind. Moreover, their reactions and responses are a great deal faster than ours.

The Horse

The parts which indicate emotions in the horse are the ears, eyes, nostrils, the upper lip, the carriage of head and neck and tail, the gait, the respiration, and any audible sounds produced.

Ears

The ears of a horse are directed forward (pricked) when

the two eyes converge upon an object situated ahead. This happens principally when the head is raised as the horse sees very little in front, clearly, unless the head is carried high with the forehead perpendicular with the ground.

A sick or lazy horse allows its ears to droop and to lose their co-ordination of action, so that one ear may be semi-erect, the other drooping to one side. Lop ears are encountered in a few horses.

Ears directed backwards along the neck indicate bad temper. In a mare, with foal at heel, this appearance may herald attack, in defence of her foal.

The Eyes

The eyes of a horse may appear bright and expressive ex-specially when the animal is excited. They may seem dull and glazed during ill-health, or when the horse is tired, hungry, or dispirited.

The Nostrils

These become widely dilated during excitement or fear. They dilate when the horse is brought out of a stable into bright light for the purpose of exercise, and they open up very widely during and after a gallop. They thus indicate excitement. The same dilatation may be seen in a racehorse at the starting gate, in the stallion introduced to a mare, and whenever the respiratory rate is increased.

Upper Lip

The horse elevates the upper lip when it wishes to attract attention, as when it expects a sweet from its owner and this is not immediately forthcoming. It does this also when it tastes food it does not relish, or when it is given unpleasant medicine. Stallions frequently elevate the upper lip and display all the incisor teeth when brought out to a mare.

Head and Neck

In the horse, unhampered by martingale or bearing rein, the neck may be straight and the head carried at a medium height. In a horse, ridden or driven, and held up to the bit, the head will be carried high with both ears pricked forwards. Arching the neck indicates courage and pride.

A head held low, dragging on the bit, indicates boredom, weariness, illness, or lack of spirit.

Gait

A light gait in which the joints of the limbs are well flexed, the horse appearing to be going "on his toes," is an indication of high spirits and a desire to race off at speed.

A slow, dragging gait, with toes "daisy-cutting," is an indication of lack of spirit, poor condition or exhaustion.

Horses are usually optimistic in outlook. The horse which hangs its head habitually and drags its toes belongs to the ranks of the pessimists.

The Respiration

This varies a great deal in the horse. At rest, breathing is quiet and the nostrils do not open widely. Any sudden alarm or anything causing excitement will rapidly cause them to dilate, with increase in the respiratory rate.

The Tail

During excitement, or when ridden or driven on the bit, the root of the tail is carried high, or level with the back. In docked horses the tail may stand almost upright in moments of excitement. A tail drawn tightly down between the buttocks indicates bad temper and an inclination to lash out with the hind feet.

Sounds

The neighing of a horse may portray a whole gamut of emotions. The voice of an excited stallion is loud and terrifying and may be heard at the distance of a mile in the country.

The gentle whinny of a mare to her foal provides an extreme contrast.

Snorting is an indication of pent-up excitement or enthusiasm. Horses snort when they see or hear hounds, their ears prick, nostrils dilate and the tail is raised.

The Dog

Of all the animals below man the dog appears most capable of expressing emotions and it would be hard to say whether the dog was accepted into the home because its moods were pleasant and understandable—or if a million of years in man's company has conditioned the mind of the dog until it reflects something of man's sentimentality.

Joy in the dog is exhibited at each end of the body, while the middle piece wriggles in time with the animal's emotions. The fore-end exhibits the ear play which fits the occasion, ears pricked or alert for joy, or relaxed and laid back on the sides of the head when the dog is being caressed—when the joy is passive and not being actively manifested. According to the looseness of the skin and the coat thickness in certain breeds, the wrinkles on the dog's brow will become very expressive.

The hinder part of the body of the dog indicates joy in two ways—;

In the first of these the tail wags rapidly from side to side when the dog is pleased, or up-and-down when in a peacemaking mood, feeling guilty and expecting a scolding. When the dog is entertaining company the tail assumes the erect position so far as its size and weight permits, and this indicates pleasurable excitement. That of the bitch at the time she is most pleased to meet the dog, is carried tightly on one side of the rump.

The second way in which the hinder part of the dog exhibits joy is when it develops a marked side-to-side movement—a waggling of the rump. This is very evident in bitches and growing puppies.

The voice may express pleasure by a series of short, staccato barks, very unlike those of an angry dog, or of the suspicious watchdog. To the experienced ear the voice of a dog gives a clear indication of its feelings, even over some distance. The bored, continuous barking of the dog shut in, and lonely, is different to every other type of barking.

Many dogs express their joyful feelings by grimacing—partly drawing the lips over the incisor teeth after the manner of the human smile. This must not be confused with the drawing back of the lips which exposes the incisor and canine teeth in the snarling dog, prepared to defend itself, or to attack. Anger is shown also by the bristling up of the hair down the back of the neck and along the spine, as well as by growling. A growl is always to be treated respectfully, especially in the smaller breeds.

The enraged dog may exhibit a typical "*threat posture*" when defending its territory, or guarding a bone. The threat posture is especially evident when an intruder attempts to enter a dog's private preserve.

A small Peke absorbs prestige from its domain and will chase away the prowling Alsatian, which bolts with a great show of alarm. It drops this directly it gets out again into the open road, when it will calm down and resume its interrupted status. The Peke does not try conclusions outside its own magic circle.

Puppies play at expressing their emotions—instinct to the fore again. The four-months-old puppy struggling with a big bone, will sound much more fierce than its own father does. When the puppy meets a strange dog it does its "appeasement act." It rolls onto its back, legs in air and usually emits urine, a few drops at a time, while it lies in this position.

Some pampered and petted dogs can even raise a sob or a long-drawn moan. In *The Mind of the Dog* I have already mentioned my bull-terrier bitch, which would moan and mourn in sympathy with any depressed person. An aunt of mine, left alone with Dinah, commenced to shed tears of self-pity. The more she sobbed the more Dinah moaned and the more Dinah moaned the more my aunt wept. Later in the day the pair were discovered, both equally exhausted.

Fear in the dog causes the tail to be tightly drawn down and tucked away between the hind legs, while the eyes assume an appearance of apprehension, and the whole body trembles.

The Cat

The bodily posture of the cat indicates very clearly at any time the reaction of the animal to any circumstance. The cat always exaggerates, it can merge in a second from a quite placid creature to a spitting maniac.

The happy cat may be relaxed, or may walk stiff-legged, with back held stiff and straight and tail rigidly erect. The relaxed cat is passively content, the stiff-legged cat is actively

seeking pleasure. The placid cat looks innocent and serene. Its ambition is to slowly rub its body upon your shins then climb onto your lap and purr. A happy *female* cat, sitting on your knees or even worse, sleeping on your pillow, may commence to "sing" its thrums." It will run its lips gently and regularly over your clothing—or pillow case—for five or ten minutes at a time, making a peculiar, subdued purring noise. It will then be found that the area she has attended to in this way will be saturated with saliva. This action appears to be associated in the cat's mind with nursing kittens. I have never known a male or neutered cat behave in this way.

The startled cat stands stiffly with coat on end and its spine partly curved, ready to jump up to a safe height. The pupils contract to vertical slits. The truly enraged cat can assume the most diabolical expression and can emit bloodcurdling sounds in language unprintable. The pupils of the enraged cat may contract to two pinholes, the one above the other, the two connected by a fine slit, the whole suggestive in shape of an upright dumb bell.

The claws of the attacking cat make their appearance from out of their sheaths, spontaneously. Those of the Siamese cat are normally partly exposed and not completely sheathed. When the cat is on the defensive, or offensive, her back arches and her tail becomes rigidly erect. The hair over the whole body stands out making the cat look much larger than it is.

The Monkeys

These animals, along with the apes, come nearest in facial expression to their human counterparts.

The happy monkey does not, however, so nearly approach a smile as a dog does, but makes rapid kissing movements with its lips, accompanied by a smacking noise. Monkeys in love poke their tongues quickly in and out of their mouths.

The small monkeys are somewhat cat-like in their body

movements when angry, but when pleased they have a way of erecting the tail and adopting a rigid hind gait with their hind limbs stiff and their rumps carried as high as possible.

Monkeys can nearly all move their ear flaps and when suspicious carry them back and pull the ears in close alongside the cheeks, at the same time pulling back the lips and exposing the molar and canine teeth. This is often done in the way of bravado and is seldom so aggressive in nature as it appears to be. The beetling brows of monkeys can also produce a frown, or when drawn tightly over the forehead may produce a demoniacal appearance.

It is not so easy to interpret the feelings of monkeys from their expression, for like human beings monkeys are expert at concealing their real feelings and substituting another facial expression for the true one.

Among Other Animals

Few birds show intention to attack, apart from geese, swans and turkey cocks. Occasionally a farm-yard rooster will attack a stranger.

Geese stretch out their necks and hiss, but swans, after a little head-bowing, will fly straight at a stranger, especially when they are nesting. Some male swans will fly at an intruder from as far away as fifty yards—not a pleasant bird to bring to a stop at that distance. Turkey cocks make a great display with wings and tails, while their faces and wattles become blood red, and may even bleed from the influence of rage upon the circulation.

Occasionally a bird, the best example of which is the lapwing plover, will try to lead an intruder away from its nesting area by pretending to have a broken wing—another manifestation of instinct—certainly not purely intelligent behaviour, although the tactics are not identical in every individual, even of the same variety.

The rhinoceros, always prone to attack if disturbed at close quarters, blinks its third eyelid rapidly across the eyeball—a sign it intends to attack.

Farmyard bulls may charge without warning but frequently

go down on one or both knees, bellow and throw any available material over their backs with their horns. They then rise and go straight away into the charge. Bulls charge blindly with eyes closed but cows keep theirs open, and follow close on the heels of their intended victim.

Chapter II

WHAT LIES BEHIND BEHAVIOUR?

During the past few years it has become increasingly fashionable to adopt a cynical attitude when discussing the mentality of animals lower in the biological scale than man, and to refuse to yield an inch when confronted by an example of what would appear intelligent behaviour enacted by one of them. But even those folk who steadfastly refuse to admit distant relationship with the higher apes will usually agree that quite occasionally one of these creatures does display some evidence of possessing faint gleams of intelligence.

And yet it is agreed by all scientific observers that the behaviour of every human infant, over the age of four months, is guided by intelligence and that every normal child has some ability to think, even before it has learned to speak intelligently. But when we make an analysis of infant behaviour we find that quite a young baby smiles at its mother, and equally sweetly at a turnip impaled on a broomstick; and while the smiling reflex appears so early it is stimulated by anything in the shape of a human head, the sight of which is apparently of instinctive value to the helpless babe, because it is the emblem which stands for food and protection from the elements.

Carefully investigated, the behaviour of the average human infant seems very little, if any, more clever than that exhibited by the household cat. Perhaps this comment is a little unfair because after all the human child, not unlike the infant kangaroo, comes into the world helpless and in a very immature condition, while the cat at six weeks old is virtually independent, so long as its environment enables it to steal for a living.

If one has any wish to pose as a scientist, one must listen to an account of anything which suggests intelligent behaviour carried out by an animal, and then remark that the behaviour referred to could undoubtedly be accounted for by some much more simple explanation. Probably this might be perfectly true, but one must be very careful not to make the same reply after listening to a glowing account of the intelligence displayed by the speaker's small grandchild.

The Mind

When we speak of the "mind" we are not referring to the physical brain but to the understanding which develops within that structure. The mind has no material composition, it can be considered only as the faculty which provides us with awareness and a consciousness of a separate existence, together with some recognition of the world around us, and of the living creatures which inhabit it.

One cannot lay open a cranium and place a fingertip upon the site of the mind, yet without the brain there can be no mind, and the quality of the mind any animal possesses will run parallel with the quality of its brain tissue.

Perhaps one might be safe in asserting that there is really no such thing as mind, just as one is safe in saying there is no such thing as light, and for that matter no such thing as vision. For what we term light is in reality only an impression which the brain receives when electro-magnetic waves of a particular length and frequency are reflected from some object through the eye onto the retina, which virtually is an extension of the brain itself. The light waves stimulate the nerve cells of the retina to transmit messages along the optic nerve to nerve cells within the occipital lobe of the cerebrum. Stimulation of the retina produces in the consciousness a sensation which we term "light," something which does not really exist in the way we imagine it does from our sensations. Vision is merely the impression conveyed to the conscious mind by the occipital lobes of the brain when these are stimulated by messages received from the sensitive retina.

In a world of trees there would be no light any more than there would be sound if a tree fell. While in either case there would be an output of electro-magnetic waves, trees have no brains which could convert these into sensations.

But when our consciousness responds to an impression we term "light," to one we term "sound," to another we term "smell," and to yet another we term "touch," then the whole of these, together with some others I have not included, merge into one composite consciousness, by which the objects around us become related to our own being. It is this composite consciousness which provides awareness of our own existence and assures us that we are living creatures.

Once aware of this, you become equally aware that you possess a mouth, eyes, ears and limbs. The seat of all this awareness is the mind, which must still be regarded as a faculty of the brain rather than as a physical part of it.

The brain cells play their part somewhat similarly to that played by the valves of a television set, while the nerve fibres take the place of its wiring system. Some television sets are more efficient than others and provide a better visual impression. In the same way the brains and the minds of some animals are superior to those of others, and so the mental impression is correspondingly clearer in the former group.

The internal "works" of the television set may be compared with the convolutions of the brain and with the white matter which contains the nerve communications. The screen on which we see the picture compares with the mind. It is the part of the apparatus which impresses the picture upon our consciousness by virtue of the electro-magnetic impulses which have been conveyed along the wires of the television set, or along the nerve fibres of the brain. When the human mind has registered one set of impressions in this way, it may automatically produce a further set, slightly different from the first. Sometimes a whole series of varying impressions may make their appearance in the conscious mind.

This ability to conjure up pictures in the mind at will, is believed to be a human prerogative. Man is said to be gifted thus with imagination. If he can compare some of these pictures the one with the other he may decide that one is superior to the others in its implications, and by choosing to retain this one in his consciousness and obliterate (forget) the others he may set into operation a chosen programme of behaviour, based upon intelligent reasoning.

It is believed that few animals are gifted with imagination, or with the ability to conjure up a *series* of mental impressions and make a choice of any one of them. Such animals are said to lack that extra cerebral activity which creates fancies, which fills the human mind with imaginings, and hopes for the future.

Animals probably are aware only of what is happening to them at the present moment, with some dim recollection of the past, and no anticipation of the future.

Human beings remember the past, usually disregard, or fail to extract pleasure from the moment, but dream constantly of a future filled with rosy pictures. David Katz, in *Animals and Men* (a Pelican Book) commented on thinking very aptly in the following:—

> "Man is capable of a kind of thinking where every step is fully under control. It is most clearly seen in scientific thinking. Both the methods of thinking and the relations between the thoughts themselves are made explicit. Such thinking is insightful in the highest degree. It is characterized by the greatest objectivity and freedom from emotional influences. Needless to say we cannot suppose the chimpanzee to think in this way at all. And even in man it is the exception. The animal remains confined to the lower realms of insightful thought. Man, on the other hand is, at least in principle, able to make use of the highest levels."

Even those animals like the squirrel who stores acorns and nuts and forgets where he puts them; the mole who stores worms, and the rook who buries acorns, bread, and anything he cannot eat, in the soil wherever he happens to be standing at the time, have no knowledge of what they do it for—and if they ever find their buried stores again it will be either by smell awareness, or by sheer chance.

Evolution, with an eye to the ecological situation of the animal, has provided each one with the kind of brain best calculated to fulfil its needs and safeguard its prospect of survival.

Animals lack the advantages of education accorded to growing youngsters of the human species, and it is therefore difficult to decide upon the calibre of the brain allotted to any young animal unless one is able to compare the behaviour of the animal with that of a human adult who has been reared entirely without educational facilities.

Historical legends are grouped around children who have been suckled by she-wolves, and within the present century one or more children have been discovered in somewhat similar circumstances, and their mental condition at the time of their rescue, and their subsequent progress, has been put on record.

It would appear that the mentality of a child who survives without the love and care of parents, deprived of constant association with human beings, remains at the level of that of the animal which suckles it. The mental progress of any child varies in its rate according to the ease or difficulty wich which it establishes communication with its parents, as well as upon the degree of intelligence *they* exhibit. The child comes into the world with a brain far superior to that of a puppy but until the child reaches the age at which it can receive messages from its parents (not necessarily by means of speech), the puppy progresses much more rapidly than the child.

But, from the time contact (verbal or otherwise) becomes established between the child and its parents, the child acquires knowledge very much more quickly than the puppy could ever do. Similarly, one will often be surprised by the intelligence displayed by quite lowly types of animals when they are "adopted" by human beings and reared in association with them, from a very early age.

I feel sure that if animals possessed the gift of speech and could receive the same sort of education granted to children, we should discover that the disadvantage of the animal lies not so much in the structure or quality of its brain as in the fact that it receives no completely understandable verbal instructions, and is compelled to worry its way through life by the exercise of its instincts, coupled with such learning on that "trial and error" principle as it may be able to attain.

A rabbit, in its natural environment, needs to be clever in only a few simple ways. Its inherent impulses are sufficient in most cases to safeguard its survival. But because an animal so low in the biological scale just manages to exist by behaving in an automatic way, there is no reason to imagine that it could not be induced to exhibit far more intelligence than its fellows if it were reared in human society and acquired experience of the ways of its human foster parents.

The child, suckled by the she-wolf, would display the degree of mentality possessed by the she-wolf. If such a child were rescued, adopted by human parents, educated among other children and its mentality fostered by human association, it might forget the she-wolf who obviously loved it, and in later life become the inventor of a space ship, or of some ghastly weapon capable of destroying a whole continent by the touch of a button. There is no limit to what human cleverness may do.

The Brain

The efficiency of a machine (and for the moment we will consider the brain as a thought machine), can at no time exceed that of its component parts. In animals the brain always fails to compare with that of man, both as regards its cubic capacity in relation to the total body weight, and in the degree of complexity of the surface fissuring of its convolutions. It is this convolutional pattern which so greatly increases the functional brain area. The cranial capacity of modern man shows a surprising degree of variations, being approximately between 900 and 2,250 cubic centimeters, with an average of 1,500 cubic centimeters. It is curious, and quite contrary to general belief, that there is in reality very little correlation between brain size and intellectual ability as judged by statistical reports.

In the gorilla, a typical example of an anthropoid ape, with a cranial capacity of 650 cubic centimeters, the large brain is due to the animal's enormous body weight, and its size does not in this instance indicate any marked increase in intellectual ability. The cephalic convolution of the human brain is roughly three times as great as that present in either the gorilla or chimpanzee.

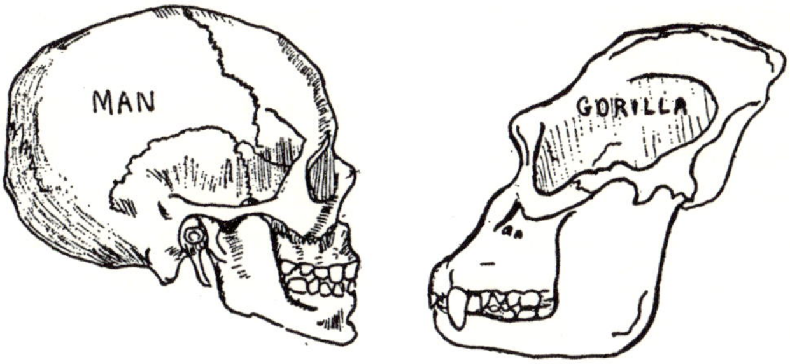

In human pigmy races the cranial capacity may not exceed 1,000-1,350 cubic centimeters.

The brain may roughly be divided into fore, mid and hind sections. The fore brain is the centre of intelligence and it is encased in the convoluted cortex—its most important part. In those animals in which sight and hearing are of more importance than general intelligence the mid and hind brain are more highly developed, and this is particularly the case in amphibians and reptiles.

In fishes, the parts of the brain normally devoted to sight and hearing are but poorly developed. In birds, the opposite is the case and in these, in which the eyes take up most of the cranial space, there is little room left for the brain itself, and even less for eye-muscles, which in some birds are almost non-existent as a consequence. In birds, too, the centres of sight and hearing are far better developed than the centres of intellect. It has even been doubted by some observers whether a great many species of fish possess that portion of the brain which other animals make use of for their visual needs. If this be the case then many fishes are not conscious of seeing, but their movements and responses are all regulated by retinal stimulation which never reaches any part of the brain capable of converting it into visual impressions of which the fish can be conscious.

The cerebellum, the hindermost portion of the brain, is especially valuable to all animals, especially birds, which depend a great deal upon the ability to balance themselves in the air and to rest perched on high upon a branch or twig.

The cortex of the brain provides the maximum surface by

virtue of folding and wrinkling in order to produce more convolutions.

The outer layer consists of grey matter, whilst the deeper white matter contains the nerve fibres which hold in communication all the various brain and sense centres. The greater the amount of surface grooving in the fore brain the higher will the animal's intelligence be. It is in this cortex that the association of ideas occurs and through this association rational thinking becomes possible in those animals which possess sufficiently well-developed "caps" of grey matter covering the cerebrum.

In some animals the cortex has been removed experimentally to determine its true function. The dog, thus mutilated, still runs away from the explosion proceeding from a shot gun but exhibits no fear of a whip or stick, since the brain, devoid of grey matter, is unable to determine any association between stick or whip and punishment. But if the whip be cracked, the dog will run away from the noise. The centre of the brain associated with fear does not lie within the new cap of grey convolutions but deep down in the old, primitive brain which lies beneath.

Loud noises cause the animal to attempt to escape, since they stimulate the auditory centre in the temporal lobe of the brain and reflexly set the limbs into operation, and so the animal tries to run away from the site of the noise. When escape is impossible the noise will cause reflex disturbances in some other part of the nervous system, apart from the limbs. The dog may vomit, develop attacks of diarrhoea, possibly with cortical disturbances which may give rise to hysteria, convulsions, incoordination of the muscles of the limbs, paralysis and loss of consciousness.

It will now become evident why veterinary services are so eagerly sought on Guy Fawkes night in Britain and will explain some of the alarming symptoms associated with the discharge of fireworks.

Different parts of the brain are concerned with different functions. The visual centres lie within the occipital lobes at the hinder part of the cerebrum, while reasoning ability is controlled mainly within the frontal lobes. Orientation, balance, and perhaps direction finding, are taken care of by the cerebellum, while the auditory centres lie in the temporal lobes.

The brain operates by means of very small charges of electric potential which travel from nerve cell to nerve cell via nerve fibres. These electric transmissions can be registered by means of the electrocephalogram. A somewhat similar instrument the electrocardiogram records the electric potential travelling through the various muscular structures of the heart wall which support the auricles and ventricles.

As I mentioned earlier, the cortex of the brain resembles a cap fitted over the old brain lying underneath it. This underlying structure represents the primitive brain of man's ancestors, the normal thought machine of the man-ape and the ape-man, millions of years ago. The original brain of prehistoric man still lies below the cortex of our own brains, and when the human being of modern times becomes overwrought, frustrated, the prey of anxiety and mentally sick, the primitive brain lying deep in man's new brain is liable to come into its own again, and take charge of the situation, whenever the cortex of the modern brain loses control, if only momentarily.

In quite a lot of human males and not a few females, loose in the world today, the cortical cap of grey matter would seem to be only a thin veneer.

In carnivorous animals there may be great variations in body size, in a weasel for example as compared with a lion, but in all animals of this class one finds relatively large brains and very little waste of cranial space to provide extensive air sinuses such as are present in most of the ungulates, especially those which have horns or antlers. The horse, which carries no horns, possesses spacious frontal and maxillary sinuses which leave room for only a comparatively small brain.

In the carnivora, the short-faced breeds of dog (the brachycephalic dogs) the Pekingese, bulldog and pug for examples, possess brains considerably larger in comparison with their body weights than the long-faced, more primitive wolf-like breeds. Possibly the short-faced mutations, unable any longer to seize their prey, attached themselves to mankind with an eye to receiving protection and free meals.

Cats, possessing even greater intelligence, adopted similar tactics with great advantage to themselves and although they condescended to admit men into their company they still retained complete independence, since they remained natural killers and kept their agility, and on the whole were even better equipped for existing under a great variety of conditions than man himself. These advantages did not, however, lessen their appreciation of warmth and comfort, a home to shelter in, and a pleasant addition to the menu.

In women the brain weighs slightly less than it does in man, but this does not imply that the female is lacking in intelligence since her grey matter is definitely superior in quality. It is generally acknowledged that her intellectual acuity is greater than that of the male, while her powers of intuition are often uncanny. It must be admitted, however, that woman's brain is not always so capable as man's of interpreting facts in their logical order of importance, nor is she always so capable of assessing their significance.

It must be remembered that woman has only recently proved that she is no longer so dependent upon man as was formerly believed, and she is now not only gaining experience but is also developing intellectually at a faster rate than appears to be the case in the male. So evident is this becoming that in some countries one may already observe a slide towards matriarchy.

It is likely that women occasionally appear less logical than men in their deliberations, not through any lack of intelligence, but because they are influenced by their hormones to such an extent that they become ruled largely by their emotions, and lose the ability to reason clearly. It must be understood, however, that individual differences in this respect may be encountered in either sex.

Among the lower orders of animals the females are less handicapped than women are by their emotions and they are almost invariably more intelligent than the males. In fact, among these lesser creatures it is the male, more often than the female, who becomes led away by his emotions while the female displays the caution and reserve. This is as it should be, since in the universal scheme the female is the important member in the perpetuation of her species, and a large proportion of males are redundant.

It is only when the brain development becomes excessive, as it has done undoubtedly in the human race, that sentiment, aesthetic values and the emotions, are permitted to influence feminine judgement.

Thinking

Before we proceed further let us consider the matter of thinking and decide what is "thought."

The dictionary tells us that we think when we turn things over in our minds. There is some truth in this, particularly when the dictionary uses the plural, "things." One seldom considers a single mental image for long at a time. The mind is imaginative in that it conjures up a succession of images for its owner to consider, accept or reject, and the acceptance of one image quickly causes a complementary image to appear on the mental screen. Presented with a pair of such images together, or in succession, one considers each in turn and comes to a decision as to which seems the better, or the more pleasant to contemplate, or more related to personal welfare or safety, or recreation.

It is possible of course that something rather similar may occur in the minds of some species of animal but how clearly one cannot tell. Nor does one know whether the animal is presented with more than a single impression, and if it has any selective choice upon which to found some rational thinking.

We ourselves do not need words in order to think on the above lines. In all probability the ability is limited to a comparatively few species of animal since only man and the higher primates are provided with the complex nervous machinery which it is believed is essential to rational thinking and a degree of higher intelligence. Even about this we cannot be sure either. We can only be guided by the behaviour of the animal and in a great many cases this indicates the possession of more reasoning ability than is usually accorded it.

It is essential to distinguish between behaviour which depends upon intelligent reasoning, or upon intuition, and that which is essentially automatic in character and based upon instinct.

Among humans there are two kinds of thinkers:

(a) Those who reason by holding mental conversations with themselves; they think in a language. Such thought is necessarily slow, but the conclusions may be more thorough, and possibly more logical.

(b) Those who possess photographic minds and visualize past and possible happenings as mental pictures, frequently attended by a great amount of detail. These are the quick thinkers.

The majority of people employ both kinds on separate occasions with a tendency to use one more than the other.

Animals cannot think in words, so if they think at all they just depend upon visual impressions, although it appears unlikely they are able to look far back into the past in this way or a great way into the future. It is hard to decide the means by which behaviour is modified by past experiences unless the animal is able to refer back to it as a mental image. Animals cannot talk interminably with their friends about their experiences, and so these do not become impressed upon their minds as our own experiences do. Probably, too, animals pay far less attention to things which happen to them.

Memory could be a valuable asset to any animal and occasionally one experiences incidents which made one wonder how much animals do remember. We do not even know a great deal about human memory and how or where the impressions of a lifetime are stored in the brain, ready to be picked out of the mental file without hesitation, and it would therefore seem invidious to profess any great knowledge of how, or how much, animals remember. That they recollect places they have visited some time previously is indisputable. Some dogs remember one person and recognise him or her with signs of joy after a separation of years - but more often they forget!

When we, ourselves, have done our preliminary thinking by means of mental images, we feel obliged to resort to language to maintain the sequence of our thoughts. We may talk to ourselves, within ourselves; or we may make notes, use figures or drawings, not only to assist our own understanding but so that we may transmit the thoughts within our brains to the brains of other people. It is this ability to elaborate thought and convey it to the minds of others which separates man mentally from all other species.

The dog, even one living in close contact with human minds, entertains only a very limited thought process. He is unable to elaborate his findings. He may visualize a situation and make a simple decision therefrom, such as whether he will stay at home in close proximity to the kitchen door until after his mistress has finished making the meat pie (with a mental picture of possible scraps), or whether he will take a stroll down the road and ascertain whether the lady Peke at the vicarage is in any mood for mild flirtation. Neither of these two thought processes demands the use of words.

Without a language how far would the human race have developed? Imagine a complete lack of reading matter, inability to write and the only form of speech a variety of snorts, groans, screams or possibly a chuckle. Imagine the absence of an interchange of ideas, and of what is now communal knowledge.

The mental pictures conjured up by the animal mind suffice for the moment without undue elaboration, and the moment is all-sufficient. Few, in any, of the lower animals make conscious plans for the future or have any thoughts which demand the help of pencil and paper.

The aesthetic sense, which begets artists, is also denied to the animal, if we except the Arctic tern who lines her nest with buds and flowers, and the goldfinch who on occasion gathers forget-me-nots for the same purpose. It is unlikely, however, that these, or the Bower bird, which uses milk bottle tops, silver paper and ice cream cartons for the nest building, have any artistic perception.

Intuition differs from thought in that it permits one on occasion to perceive the truth without reasoning or analysis. Such inspired knowledge makes its appearance in some of the lower animals as much as in man. It has been said that the writers of the Bible were inspired by divine influence, which is probably another way of saying they possessed intuition. Not only child prodigies are born with intuition but also sheepdog puppies, which begin collecting sheep from a wild moorland country during the first year of life and shepherding them home through narrow gateways regardless of the inclinations of the sheep. The members of the sheepdog breeds are imbued with an inherent consciousness (not based on experience) of exactly what passes through the mind of the sheep, together with the ability to forestall its impending action before the sheep can put it into operation.

From my own long experience I am convinced that intuition plays an important part in the behaviour of a great many kinds of animal, probably with more regularity and greater accuracy than in man, who is far more likely to regard his intuition as "just a hunch," illogical, and therefore of doubtful value. In consequence, he substitutes for it some more reasonable type of behaviour which often proves far less successful.

One of the things which has always struck me forcibly is that the animals which man regards with contempt appear to me to be the most intelligent. It is common to term a person as ass, or a goat, or even a goose, when in point of fact these three animals rank very highly among their fellows so far as intelligence is concerned.

A good many psychologists believe that animal behaviour follows a set pattern in that the conclusion of one action, or series of actions, acts as the stimulus which sets the next series of actions into operation. If this be true then every animal, excepting man of course, must behave in the same way as a machine which produces material in a number of stages, the completion of each manipulating the levers which start up the machinery required for the next stage.

It is hard to visualize an intelligent animal as being automatic in its behaviour without free-will to change the pattern when it so desires, and many of us who have spent our lives among animals in the natural state do not accept this theory without a certain amount of reserve.

Instinct and Needs

The man in the street has a simple explanation of the cause of all the seemingly inexplicable habits of animals. He provides this in one word - "*instinct*"! During recent years it has become quite unfashionable to mention the word in the hearing of our scientific friends. Nowadays, a number of other terms are used in its place. One speaks of needs, drives, urges, and may even employ far more complex terms. The intention is to bring the conception of instinct into narrower channels. Needs, as one may gather from the word itself, spring from within the animal and produce an active response. Animals have a need to survive which they satisfy either by flight or fight. They need mates, the society of other animals of their kind, and young animals in particular have a need of recreation - play devoted to growth and acquiring experience of combat and escape.

Needs account for only a part of an animal's instinctive responses. Hunger is a need; shelter, warmth or coolness, may be others.

But needs are not constant and not always obviously present. The need to hibernate, the need to mate or to migrate all have a seasonal urge. They may exist but they remain dormant until a further stimulus, usually the presence of a hormone in the blood stream, stirs them into activity. Needs may be innate, or ready to exhibit themselves immediately after birth. The need to suckle is present in all freshly born mammals, and it may be exhibited in most babies, whether they be those of lionesses, sows or women, by the lip smacking noises the infants make before they even find the teat or nipple.

All of us at some time or other have displayed a need for

milk, but we may later develop needs for other things not included in the scheme of nature. We may experience a need in relation to alcohol, tea, orangeade or tobacco. None of these could be readily regarded as instincts any more than a need, which has become a craving, for heroin, cocaine or morphia, could be considered as such.

Needs in the psychological sense are concerned with necessities. Luxuries do not enter into the lives of animals in the wild state, but in the human race the luxuries of yesterday become the needs of today.

We may experience a *drive* in a number of directions. People declare that through domestic unhappiness or for some other reason they have been driven to drink, or immorality, or even to attempting suicide. None of these activities have been activated purely by instinct since they are not innate urges handed down from ancestors, who had probably never tasted alcohol, indulged in immorality, or held their heads in gas ovens.

Perhaps some of these new terms are applicable to human psychology rather than to that concerned with animals. I think we shall introduce fewer complications if we keep to the use of the term instinct for the remainder of this volume.

David Katz, in *Animals and Men,* (Pelican series) gives two definitions of instinct. The first of these is G. E. Müller's modification of that given by William McDougal. According to Müller an instinct is "a psycho-physical disposition dependent upon heredity, very often completely formed immediately after birth, at other times only after a certain period of development - a disposition which leads the animals to pay particular attention to objects of a certain kind or in a particular way, and having perceived them, to feel an urge towards a definite activity in connection with them."

The more one sees of animals in all their varieties and patterns of living, whether they be parrots or politicians, the more one perceives that the greater portion of the intellectual life of each individual is spent within the limits of a circle, the radius of which is determined by the nature of the instincts possessed by the particular animal.

Every individual experiences great difficulty in passing outside this circle and among humans it is only the born rebel or individualist who ventures on short tentative excursions beyond its circumference. Such diversions beyond the accepted territory demand the exercise of will power - the spirit of adventure - and they may be attended by serious misgiving, the experimenter endeavouring to persuade himself that conservatism is not always the best policy and that a rebellion against one's inherent nature may bring rewards, well worthy of any risk entailed.

Few animals below mankind entertain such rash inclinations, partly because they are devoid of imagination and partly because such enterprise lies outside the range of their instinctive impulses. While man has advanced and modified his behaviour in accordance with the experience gained outside the limits of his instinctive circumference, the lesser animals have kept strictly within traditional bounds, with the result that in all animals other than man, still living in their natural environments, their behaviour follows closely the same pattern that was in vogue a million years ago.

The second definition of instinct exemplifying the more biological type, given by David Katz runs:

> "Instincts are complex groups of conditioned acts which, though they contribute to experience are, on their first occurrence, not determined by individual experience; which are adaptive and tend to the well-being of the individual and the preservation of the race; and which are similarly performed by all members of the same more or less restricted group of animals."

A baby thrush comes into the world fully equipped with the elements of craftsmanship which enables it, when sufficiently strong, to crack a snail shell by holding its edge in its beak and beating the shell upon an anvil such as a stone. It has never been taught to do this, nor does it learn by mimicry.

Away back in the days of our primitive ancestors no child came into the world already equipped with a knowledge of the easiest way to skin a rabbit - or whatever in those days took its place. It learned the art later by watching and imitating its parents. But a rabbit must first be caught - something beyond the power of a young child - but any child may pick up a nut in the right season. If it had sufficient experience to know that a nut contained a kernel it would not be long before a child of five years or more would be dropping a stone upon it, or exercising its intelligence in some other way by which it might reach the contents.

The child does none of these things by instinct, but by intelligence and intuition. Few children will set about nutcracking in precisely the same way, but one young thrush will hold the snail shell and crack it on a stone in exactly the same fashion as every other young thrush - not from the use of intelligence but purely by the prompting of instinct.

But the young blackbird does not possess this ability to crack a snail, either when it comes into the world or afterwards, so it remains in hiding and listens. When it calculates from the sound of the shell hitting the anvil that the precise moment has arrived, it divebombs the industrious thrush, grabs the naked snail, and carries it away to its young.

This trick seems to be based upon intelligence and experience, but it may be as much the outcome of instinct as the behaviour of the thrush it robs.

The same cute blackbird, later in life, in a dry, early summer, unable to find worms, or rob a thrush of its snail, will spend ten minutes vainly attempting to push a large, dry crust down the throat of its unwilling youngster - an example of the employment of an instinct to no purpose, simply because the blackbird is following its impulse without tempering it by intelligent thought.

If we listen to the teachings of Freud we shall discover that man is the victim of a great many instincts of which he is not conscious. Human beings do retain a number of instinctive forms of behaviour, but living in a so-called civilised world, they may be compelled to repress a great many of their normal urges. As a result, the human race suffers from a universal neurosis — something almost unknown to animals which lead natural lives.

It is when an animal is called upon to perform some action, extricate itself from a difficulty, or carry out some type of behaviour which could never have been experienced by its progenitors, that it may show evidence of an ability to reason, quite uninfluenced by inherent urges.

An example that comes to mind occurred in connection with an aged pony mare, turned out to graze, as a pensioner, for the first time in her life. She contrived to pass one forefoot between two tightly stretched strands of barbed wire, something she had never before seen. Instead of going into a panic as so many horses would have done with the infliction of serious injuries as a consequence, the mare stood perfectly still for a whole minute while she contemplated the situation. Then she slowly advanced the hind foot of the same side, inserted the toe between the two strands of wire, pressed down upon the lower, and carefully extracted her fore foot.

One might reason too that the young mouse which finds a trap baited with cheese, tied in place by loops of cotton, lies on its side and very gently nibbles through each strand, liberates the bait and carries it away without setting off the trap, is displaying intelligence (and bravado) not based upon instinct.

Animals denied education and the spoken word are less impressed by happenings than the child, and their memories are not so keen. It is equally improbable that a child would retain its memories were it not for its association with older people. A grazed knee would seem a trivial matter to a child or to a puppy, and after the first pain and shock had passed would soon be forgotten. But when the knee of a child is washed at frequent intervals causing fresh pain on each occasion, is fussed over, disinfected and bandaged, when it necessitates an injection of tetanus antitoxin and two days in bed, and especially if the incident is discussed over and over again in the child's presence by worried parents and relatives, the memory is likely to be retained for a very long time. Similar attention would make an impression upon the mind of a puppy, or any other young animal.

The difference between the child and the puppy lies in the fact that the former develops a complex linked up with its injury. It becomes interested in the treatment, and revels in being the centre of attraction, if only for a short period.

The puppy would be aware that it had a sore knee providing that the injury caused it pain. It would also know that at intervals interfering humans approached it, held it rather too firmly, and wrapped cloths around its leg, which it removed as soon as the opportunity arose. But never for one moment would it realise that the treatment it was receiving was in any way related to its sore knee, or that it was being carried out with kindly intention.

Instinctive behaviour in animals may appear in many different forms. When it assumes complete control, as it does in most insects for example, and in many of the larger animals which live in flocks and herds, all individuality is lost and the animal becomes just one unit in a multitude. Normally, the behaviour exemplified by the unit is adapted to the needs of the one species in one type of environment, and when the environment changes, or the circumstances normally associated with ecological conditions become varied, the instinct may fail in the purpose for which it was originally intended, and may actually prove a menace, in that it may expose the animal to perils with which it is no longer able to contend. When the change in circumstances involves the whole species there must unavoidably be a depletion in the number of its individuals. But a few of these individuals out of the multitude may display some degree of elasticity of this instinctive urge, and this may permit the few to vary their behaviour in some degree, so that it now adapts itself to cope with the new conditions prevailing. Alternatively, certain units may prove to be in some way different, physically or mentally from the rank and file, so that they are in this way better able to meet the new situation. As a result there will now be some survivors, and if these pass on their own special characteristics to their progeny the species will again multiply until some other change in circumstances brings about a repetition of near-disaster.

It is not always this elasticity of instinct which enables a species to survive. Sometimes the factor which introduces a lethal risk is disease. But even in the fight against this evil, instinct may play a part in survival of the species, although it may be an instinct possessed by the lucky few.

When myxomatosis almost wiped out our rabbit population, the disease was spread by fleas, and these were present in great numbers in the burrows underground. Most female rabbits either gave birth to their young ones in a deserted burrow, or in one they had dug specially, well away from the main burrow. The purpose of this was to escape the attentions of the prowling buck whose instinctive behaviour caused him to turn his back upon the nest and its helpless infants and kick them into eternity.

Some does, finding that even these measures did not hide them safely from the philandering male at this critical period, delivered their young ones in nests scooped out on the surface of the ground within bracken, or other cover. Although this exposed the litters to birds of prey, foxes, badgers, stoats and weasels, the risk from these creatures was not so great as that from the buck rabbit, partly because the instincts of marauding animals led them to seek young rabbits *underground.* Because the litter remained on the surface (thus adopting the behaviour of the hare which, although susceptible to the disease largely escaped it as a result of its instinctive behaviour), the young rabbits did not come into contact with infected fleas and so the rabbit is still with us, even if in reduced numbers. But while the instinct to avoid burrows (and incidentally the rabbit flea) remains, so their numbers will multiply.

It may well be that the instincts of the rabbit family contained this degree of elasticity long before myxomatosis swept down upon them, and that at all times a small percentage of does *did* prefer the surface of the ground to the burrow. Even so, this only goes to prove that a complete rigidity of instinctive behaviour can have its limitations of service, and that in the long run a little individuality often pays good dividends.

Another of the disadvantages of strict rigidity of instinctive behaviour pattern is that the individuals of the species never advance their education. Learning is usually acquired in the animal world on the principle of trial and error. When the pattern of behaviour is set and unchangeable there can be no experimental, and seldom any accidental, variation. Accordingly, the animal makes no trial of fresh methods or patterns, and never profits by its error, presuming it survives it.

The whole life of the bee, for instance, is controlled by a series of instinctive acts, the cessation of one providing the stimulus for the next, although (as in the normal reproductive behaviour of mammals), a hormone, or some chemical substance taken in with the food, conditions its whole course. The bee has no use for trial and error. Its present programme fulfils its purpose most admirably and could not be greatly improved upon.

In the animals below man, the behaviour influenced by a combination of instinct and hormones needs no variation either. In every species it is so arranged that each stage of the journey from meeting and courtship to the final farewell to the offspring, is organised on the principle of a world tour in which the completion of one stage of the itinerary produces the tickets and instructions for the next.

This does not apply in quite the same way to human courtship, for although both men and women have their instincts and their hormones, their extraordinary intelligence enables them to vary the pattern of their love life until it becomes hardly recognisable. Animals do not suffer any mental agony or emotional upheaval as a part of their reproductive cycle since it has for them only a short, seasonal importance, and once over is quickly forgotten.

Kinds of Instinct

We may regard the natural inherited instincts of animals as *automatic instincts.* As Müller pointed out in his definition, an insect may be completely formed within a short time of leaving the egg, or it may go through a series of changes, and at each stopping place it may have its own special instincts. These are often termed *learned* or *acquired instincts,* since they have been cultivated as the necessity arose, or shall we say that only those individuals which had learned or acquired particular patterns of behaviour at each development stage, survived. In each of these stages the instinct which produced the behaviour was in reality present at the time of birth, and in fact travelled through the egg from its parents.

The stimulus which induces instinctive behaviour may be an *internal* one, conditioned perhaps by some hormonal secretion from an endocrine gland, or by some sensation arising from some part of the body. Alternatively, the stimulus may have an *external* origin. This seems the best place to mention what is termed a *reflex,* which gives rise to *reflex action.*

A simple reflex is illustrated by the muscular contraction produced, let us say in a limb, when the skin is pricked by a needle. This is merely an orthodox instance - the reflex actions produced every day in all of us may present an endless variety.

But a reflex may not be quite so simple. The stimulus may be a visual one - as when our eyes first detected the bull in motion - and the reflex actions included not one set of muscles but almost the whole of the muscles of the body, all operating in ordered sequence to produce a co-ordination of bodily movement, which landed us on the right side of the gate. In this instance the sight of the bull acted as the external stimulus. In another it might be an auditory stimulus, or it could even be a sensory one connected with taste, as when one takes a mouthful of scalding hot tea and promptly ejects it quite automatically, whatever company may be present to exert an inhibitory influence upon such behaviour.

Instinctive action differs from reflex action since it follows a set pattern and is automatic in nature. It proceeds always from *within* the animal, and is a manifestation of some part of the normal behaviour of the animal's ancestors handed down to the progeny. It is designed to *anticipate* possible happenings and provide a remedy, and it may come into operation, when the occasion demands it, without any external stimulus.

Reflexes operate retrospectively and not in anticipation of an event, and the impulsive action to which they give rise is never behaviour which has been contemplated and organised in advance. There is a difference in the cause which lies behind the act of running away from a bull, and of running equally fast and jumping a hurdle at a sports meeting. The one is impulsive and entirely automatic, the other well practised and elaborated in anticipation.

The higher the biological status of the animal the more elaborate is its behaviour, but the *instinctive* behaviour of a bee or an ant, or of some of the creatures which bore upwards through inches or feet of solid earth to emerge at the exact spot where food lies on the surface, goes far beyond any of the instinctive behaviour still retained by man himself. Man now depends upon his instincts very little. He acts reasonably - at least when he is in the mood to do so.

An instinct may undergo change, and develop as the young animal grows and matures. Although the form it takes at each stage of growth was already dormant in the animal's brain at the time of birth, it requires the action of a hormone, derived from one or more endocrine glands, to pull the trigger that sets the instinct into operation at the right time.

Thus, a puppy up to six or eight months old, urinates with all four feet on the ground, crouching slightly, with its hind legs wide apart. If it is castrated as a puppy it will in all probability retain this method throughout life. As a puppy it will be able to empty its bladder, in one effort, but later, as it attains puberty, it will cock a leg at every available post, or whenever the smell of urine of another dog or bitch is manifest. The intention is to advertise its presence, and the underlying motive (unknown to the dog) is directed towards subsequent mating and reproduction. The inclination to urinate sometimes on grass, or leaves, and then to kick the soil or leaves in all directions with all four feet, appears to have a double purpose, the one to widespread the odour of the dog's own urine, the other to anoint his own body with sufficient of it to impress its character upon all other dogs or bitches who may wish to include it in their catalogue.

The instinct to guard the owner's premises, which, incidentally, are the dog's own, is probably linked up with the preservation of territory common to most animals, and is one conditioned by the absorption of testicular hormones.

Salmon and other fish, which enter rivers to spawn, are also conditioned in their behaviour by hormones, although this does not explain their ability to return to the particular river in which they themselves were spawned.

The nest-building of birds, the egg-laying and the broodiness, followed by the maternal instinct to feed and keep warm the young birds, comprise a number of stages in the reproductive cycle, each of which is controlled by a separate hormone.

Even the hunting wasp has carried out precisely the same pattern of performance whether the year be 1960 B.C. or 1960 A.D.; time makes not the slightest difference to the ritual enacted by every wasp of its kind. The wasp makes its nest, stings and thereby injects a tranquiliser into the body of a spider or caterpillar, and so renders it unable to escape. It lays its eggs upon its victim, seals the eggs, and the future meal of its young ones, inside its nest; then it moves off to other business, knowing nothing and caring less about the result of its labours and entirely ignorant of their purpose.

When I mentioned the fact that the nest-building, egg-laying and broodiness of birds was conditioned by hormones, I made no mention of the fact that all the art and craft associated with building the nest (and there are countless kinds of these) was inherent in the brain of the fledgling and that all that the hormone does is bring this knowledge to the surface. The bird employs it to build the nest without knowing why it does it, what it does it for, or what it will do after it has done it.

By the time the nest is finished the henbird has developed a warm spot on each side of the breast, and from these the feathers are rapidly disappearing. To ease the discomfort of these "brood spots" she sits in the cold nest until she discovers something warm and solid beneath her. The cockbird takes a look, becomes exceedingly excited, and announces to all his kind by his noisy chattering that his mate has laid an egg. Masculine excitement concerning the arrival of an heir is not unknown in the human species. The male bird becomes quite exhilirated over the arrival of the egg, which is an earlier stage in the reproductive cycle than birth. If the father bird happens to be a gull, he will guard the egg with his life while the hen is away and the egg is still in the nest, but if by any chance it should accidentally roll outside its magic circle he will pounce down upon it and eat it with gusto. An egg *inside* the nest is a stimulus which brings the paternal protection into operation. An egg *outside* the nest is an entirely different proposition. It no longer provides the stimulus, and to the gull it then represents a very tasty snack.

Harking back to nest-building, can one imagine a young woman, or a young wife, going straight into a factory, and without any tuition whatever picking up a bundle of osiers and converting them into a perfect baby's cradle? This would appear very unlikely, and yet a cradle is not nearly so complicated a structure as a nest of a thrush, or a wren. A female spider will weave a web of the most intricate geometrical pattern, but our young housewife would be incapable of producing a pattern of lace, or weaving a carpet, until she had received at least twelve months' tuition. Not every young woman would be successful then, but *every* young female spider carries out an equally difficult task without any teaching whatever.

Baby chickens are not expected to display advanced sexual behaviour, and yet if a chicken, two days old, be injected with testosterone, a derivative of the testicles, it will shortly afterwards attempt to "tread" its mother, or some of the other chickens accompanying it. It will start at once to grow a comb, and by the time it is a week old it will be commencing to crow!

A young garden spider just out of the egg will in all probability devour one or two of its brothers and sisters to give it table, it will at once commence to travel in circles and then set strength. Removed from its repast and dropped onto the office about spinning itself a web. If its first attempt holds no promise of capturing a fly, it will find a corner or angle where it can produce a web of geometrical pattern equal in style and finish to any web spun by its own very experienced mother.

There is no hormone at work here as far as one can ascertain, but simply an instinct shared by every garden spider. Not every kind of spider, though, since quite a number hunt just as a fox might do, while others set ingenious traps to capture their prey, traps which are quite as cleverly produced in their own fashion, as the garden spider's web.

A spider of any age, touched with a feather or pursued by a shadow produced by holding a pencil between it and the source of illumination, will at once roll itself into a ball and appear to be dead. It will stay motionless and feign death for several minutes at a time, if necessary. An opossum will do precisely the same thing, and so will a good many small animals and some kinds of birds.

Partridge chicks, running behind their mother will "freeze" in this way if a hawk comes into sight, not because they see the hawk but because the sight of a hawk releases an instinct in the mother partridge which induces her to utter one note. This "hawk alarm" note is common to a great variety of small birds and may be regarded as a universal signal. As soon as the danger is past a second note, in a slightly different tone, gives the all clear signal.

In a good many kinds of ungulates, sheep, deer, and the like, as well as in pigs and wild ponies, the sight of a stranger or of some unexplained object, will produce a reflex grunt. This is a true reflex and is not an instinctive response.

The visual (or sometimes auditory) impression acts as a stimulus and produces a sudden contraction of the muscles lying between the ribs, the muscular pillars of the diaphragm, and of the muscles of the belly wall. The contraction expels air from the chest through the trachea and larynx, and causes a vibration of the vocal cords. This signal is recognised by all the above animals and is in itself often sufficient stimulus to reproduce the reflex in other members of the herd of flock, so that the sound travels over a distance and gives warning to all the herd members within the locality.

Baby kangaroos are born after only a few weeks gestation and achieve their development within the mother's pouch. They come into the world little larger than a fat slug and have to at once negotiate a journey from the genital orifice of the mother into her pouch, a distance of possible six inches or more. They have small arms and "hands" at this stage of their development but no serviceable hind limbs. It was believed that the mother kangaroo, who was quite well formed "hands," lifted the baby into the pouch, but it is now known that she merely licks a path through her hair and expects the infant to make the journey under its own power.

The secret lies in the instinct granted to the immature baby kangaroo. It could be almost summed up in the one word, "Excelsior." The ambition of the youngster is to climb higher all the time. The mother therefore settles herself with her back against a tree trunk or other convenient rest and steers the infant toward her pouch by elevating the appropriate portion of her body.

As soon as it is safely in her pouch a further instinct induces the baby to grip a teat in its mouth, and having once secured this, it continues to hold on for days, or weeks, thus anchoring itself in a secure position.

Among many of the birds of prey, the cockbird brings in the game and the hen tears it up into pieces of suitable size to feed the young. If by any mischance the mother is lost, the cock still brings in the provender but has no instinct to dismember it. As a result the youngsters, unable to tear up a rabbit for themselves, often succumb.

Occasionally instincts are at fault in individual cases. For instance grey rabbits will "freeze" in their tracks and be unnoticeable crouching in a wheelrut, or in a tuft of dry grass. White rabbits not uncommonly appear among ordinary grey rabbits, either as sports or as introduction from tame stock. These possess the same "freezing" instincts as the grey rabbits and instead of becoming inconspicuous, stand out clearly against their background. This applies equally to white partridges and white blackbirds. The latter possess the instinct of all the species to seek shelter in holly trees in gardens and hedgerows. Against the dark foliage they can be clearly seen.

Some varieties of instinctive behaviour seem to have no definite purpose. Processional caterpillars will follow each other around a circle of wood or around an old motor tyre for days on end if they are collected and set upon their way.

A hundred, or a thousand lemmings (sometimes a hundred thousand) will start out on trek, marching through or over every obstacle in their path, until they walk into the sea and drown, or fall over the edge of a cliff. Probably hunger is often the stimulus which sets the column in motion, but an instinct to travel ever onward must be a destructive one, unless as in bird migration it has a definite purpose in view, even if the purpose is not apparent to the animals participating in the migration.

One could recount innumerable instances of instinct in animals but the foregoing illustrate most of the types we have discussed. It is interesting to note that in some forms of plant life one may encounter behaviour which should be considered instinctive, sometimes intelligent, if it were carried out by animals.

Many of the plants set traps to capture insects fall into this category and one can draw only a fine line between some of these and creatures such as the sea anemones and the like, so far as their visible behaviour is concerned.

In practically all kinds of animals the number of responses made to any specific stimulus, within a given time, may not be unlimited, provided that the stimuli are repeated at frequent intervals.

A simple example of this is seen in the behaviour of a garden spider; although one may carry out experiments of a somewhat similar nature on a great many of the larger animals with similar conclusions. If on a fine, bright day, when the spider is in residence and the web dry, you take a long wax taper and use it to gently vibrate some part of the web you will find that the spider will rush out of its lair in search of a fly entangled in the web. As a spider has a visual range of only about six inches, you need take no pains to keep out of sight. If you repeat the vibration at intervals of a minute you will discover that for the first four or five times the spider will respond readily. When it finds nothing in the web its enthusiasm commences to fade, until after about the eighth or ninth attempt you may discover that the stimulus produces no further response.

The male spider (who may be eaten by the female unless he displays great caution) makes use of this failure on the part of the female to respond to repeated stimuli, when he sets out on a courting expedition. He first tweaks the web and retreats something after the manner of a schoolboy ringing a doorbell. When this produces no further response he commences a series of semaphore signals with the aid of his palps. When the lady has tired of these, he produces his gift, possibly a fly, wrapped in many layers of web. While the female unwraps this the male may complete his mating programme - or the female may decide to put another of her instincts into operation. This one impels her to eat the first two or three of her bridegrooms to satisfy a need. Spider or insect juices are essential to the development of her full fertility. This commonly happens among insects, too (the spider, of course is not an insect), as is exemplified by the blowflies, which need at least one good feed of blood, or meat juice, to ensure fertility of their eggs. The female mosquito sucks blood for the same reason, something the male does not do.

It must be understood of course that the spider, in company with the other blood-sucking insects I mentioned, has no knowledge of why she requires to satisfy this need.

While instincts may or may not arouse pleasurable sensations, it appears that the satisfaction of most needs may do so. The relief of hunger, thirst, and the operations of defaecation and urination produce feelings of satisfaction, and it may be presumed that the ingestion of a bridegroom may be equally comforting to the spider. The pleasure derived from the gratification of a need frequently ensures that its importance is not overlooked.

Tropism

A *tropism* is a reaction which embraces the whole of the body in relation to a stimulus.

The stimulus may be light, which attracts many animals to approach its source. This applies to moths and men, as well as to most of the animals between these two extremes. Most insects fly towards the light, which is why they favour windows to walls of a room, and it is well known that although we cannot approach much nearer to the sun, itself, a great many animals enjoy basking in its rays. At the same time there is a very large population of night-loving animals and among the insects, the worms and many of the rodents, darkness seems to be favoured, largely perhaps because it provides greater security.

We have briefly discussed the phenomena of claustrophobia and agarophobia in Chapter One.

Tropisms affect all living forms from bacteria and algae up to the highest types of life. One of these tropisms is characterised by an attraction towards certain chemical substances. The fascination which some scents and odours have for human beings and dogs, to take two animals for consideration out of many, may amount almost to a tropism where the olfactory sense produces a general bodily reaction which impels the animal to seek the source of the scent. Bees, butterflies and moths are also greatly attracted by scents and their life history and their reproductive behaviour are largely linked up with their olfactory perceptivity.

A great many animals find the earth very attractive and feel an urge to dive into it - or they may endeavour to squeeze themselves beneath stones or other solid substances, and hide in dark places if a hole in the ground is not available. Similarly, others like to clamber up high mountains, or to find refuge in the treetops. The birds are unique in that they have the choice of the open air or the shelter of trees, or they can share the darkest caves with the bats, according to their own special proclivities.

Escape Reaction

A very important instinct, more commonly observed in wild animals and in human children (in which it is well-marked), is one directed towards survival of the individual. It is known as the "escape reaction." It is characterised by an impulse to run away when approached by another animal. In a natural environment wild animals each have their own approximate escape distance, which can most easily be expressed in the number of yards below which they will refuse to be approached by animals of another species.

The animal approaching is usually one recognisable as an enemy, which implies that it is liable to attack or kill if it is permitted to approach sufficiently near. On the other hand, it may simply mean a strange animal, regarding the character of which the other may have no knowledge or experience.

This urge to escape becomes lessened among animals in a state of domestication. In children it diminishes as the child grows older but usually persists into the seventh year of life, though in many children it may remain in evidence a great deal longer.

Horses, in the wild state, unaccustomed to the sight of men, will bolt as soon as the intruder gets to within 150 yards, especially if the human scent is carried towards them in the wind. Moorland ponies used to traffic and to being fed by visitors, may make no effort to escape, and may even approach visitors in the hope that they carry food, or tit bits.

Horses used for riding or driving, turned out to graze, will frequently establish an escape distance of five yards, unless the person carrying the halter is also provided with a tin containing oats, or something equally attractive.

A lion, unconcealed, can be approached by a man up to a distance of 25 yards before it attempts to move off, unless it happens to be hungry, when it may approach instead.

Wild animals which see men frequently, often modify their escape reaction to accord with their experience of the behaviour of the species.

Wild ungulates, deer, antelope and other similarly fast-moving animals, will allow a lion to come within 50 yards before they make any attempt to bolt, but they seem often to possess an uncanny knowledge of whether the lion is hungry and out for a kill, or whether it is taking a stroll. They will, therefore, vary their escape distance to meet the situation.

In almost every large herd of small deer or antelopes, one or more will be found which do not demonstrate the escape reaction in time to avoid the lion. Why this should be is not very clear unless the animal is parasitised and weak, or suffering from some disease. Capture by the lion precludes all possibility of making a post-mortem examination of the carcase.

In the jungle wild animals usually move off as soon as they become aware of the smell of human beings, and this consciousness varies according to the force of the wind (if any), and its direction. Deer will scent a human being at a distance of 500 yards when the wind is blowing towards them, but in other circumstances they may be approached to within 75 yards provided they do not see or hear the stalker.

Giraffes move off when about 200 yards from the person or animal disturbing them.

A rhinoceros will sometimes make its escape when approached at any distance up to 100 - 200 yards, or it may decide to attack rather than retreat.

Some animals, from dogs to gorillas, specialise in making mock attacks in place of attempting to escape. A man gorilla will sometimes make a headlong rush as though determined to kill, but having covered half the distance between himself and the intruder will sheer off and return to his group. One can never be altogether certain whether the charge is genuine or otherwise, until the gorilla provides the answer.

Anyone who has endeavoured to capture a frightened stray dog, loose in the open, will soon discover its own special escape distance.

This act of taking to flight in animals is based upon the setting into operation of certain combinations and co-ordinations of muscular movements, similar to our own which we discussed when we were considering the means of escaping from the angry bull.

Occasionally you may wish to catch your own horse, or pony, which is grazing in a large, open field. In this case the escape reflex may not be manifested in an unqualified form, being governed by a less intense series of reflexes. The efforts on the part of the horse to escape may alternate with others intended to establish friendship and contact, though in a rather half-hearted fashion. Soon, the escape reflexes begin to give place to those associated with establishing contact.

The degree of vigour attending the escape behaviour will be tempered by the animal's experience. In most animals, however, especially in the ungulates, the instinct lies dormant even after experience has proved that the human being is not one to be dreaded. It is due to the persistence of this primitive instinct that horses sometimes run away from their owners when left unattended for a brief second, or when they have parted company at a jump and the rider has neglected to "hold on to the reins."

The instinct may be so deeply instilled in the horse's mind that it may "bolt," even when being ridden, from objects which to our eyes would appear completely innocent and unlikely to cause alarm. This may indeed be the case but instincts will not be denied for ever, and frustration may produce odd results, which is why caged animals and those within rails, so often make a frantic effort to escape, and in doing so cause extensive injuries to themselves.

"Shying" in horses is also due to a reflex which operates when the horse is suddenly startled, or sees something it does not understand, and feels that the escape distance itself and the object is insufficient. After the swerve, confidence may be restored, or the horse may decide to bolt in spite of every attempt on the part of the rider to restore confidence and regain control.

Every fit and healthy animal is at all times ready to seek safety in flight. Fear is a stimulus likely to make its appearance for very small reason, which is why the child sent upstairs to fetch some small article, especially during darkness, will suddenly turn tail and rush down again minus the article. It is also why little girls giggle, which is *"substitute behaviour"* taking the place of running away from the person speaking to them. Most children up to six years of age put the escape distance into operation at the approach of a stranger, especially in country districts. Town children become conditioned more readily and do not bolt so frequently when they are asked the way to the railway station. When two children are together the escape reaction less seldom becomes manifest.

The escape distance is not altogether constant in any animal and it may vary among wild animals in accordance with the nature of their environment, and also with the degree of experience of intruders possessed by the animals. Most of these have some pre-conceived notion of the direction of their flight should it become necessary, and here again a tropism plays some part since animals usually make for light or for darkness, for the trees or the depths of the earth, according to their kind.

A rabbit, disturbed by a dog or a fox, does not dart across on open field if it can avoid it, but seeks a hole through a hedge and keeps as close to cover as possible. Every small animal knows each bolt hole and hiding place within its own territory and carries a picture of every potential refuge in its mind. A bird aims at the highest tree or evergreen bush; the rabbit seeks the run through dense cover which will lead it to its hole.

Animals in native reserves become used to seeing human beings, and their escape reaction becomes modified. This is more often the case when the human beings convey meat or hay. The animals are far less afraid of mechanical transport than of a man on foot. Even on our own country highways one will notice that birds and animals nowadays are hardly at all scared of our fast-moving motor traffic.

Learning in Animals

The whole of animal behaviour is not directed by instinct and even in those animals in which instinct is fully operative the resulting behaviour can often be modified, improved, or perfected, by learning.

Although the time for learning is during the stage of growing-up, animals, like men and women, continue to acquire fresh knowledge throughout the whole of their lives.

Learning is acquired in two ways:

(a) ***Trial and Error***

The animal, being its own teacher, recognises its mistakes and takes steps to remedy them - or recognises an improvement in technique and decides to adhere to it.

When animals depend upon their own intelligence to improve a technique they must come very close to rational thinking. But the change of technique may have a more simple explanation - the discovery which gives rise to the improvement may be quite accidental, but the amount of improvement may impress itself upon the mind of the animal and induce it to continue with the new and improved technique.

To take a simple example:-

A dog, living in a country house, goes for a run every morning after breakfast. On returning it enters by the front door which is always open by day. On one occasion, in order to avoid a caller and his bull-mastiff, the dog runs around to the back door and enters the house through the kitchen. The new kitchen-maid is fond of dogs, and gives him a few scraps of meat. From then onwards the dog enters always by the back door. The new behaviour was introduced by chance, but it persists because it produces a more satisfactory result.

It could be as simple as that.

(b) ***Parental Instruction***

This is the way that children learn even if at an early age the parental instruction is instilled by proxy.

All young animals do not come into the world ready to adopt the behaviour common to the species. The instinct is there but before it can be put into operation the youngster requires instruction.

It has to learn the rules of the game. Young otters, for example, do not take naturally to an aquatic life, and it is likely that if one, taken from its mother when only a few days old, were reared by hand and fed on rabbit and chicken away from all water, it might experience no pressing desire to dive and swim and catch fish. Nor does it follow that it would learn to do so in the presence of water unless it had its mother present to teach it how to set about the task. In its natural environment it has no alternative - it is taught to swim by its mother who catches hold of its tail and pulls it in, and it then holds onto its mother's tail until it has learned to keep afloat, and to take a few shallow dives. It learns to feed by taking fish from its mother's mouth. Very shortly it finds it can only obtain the fish by chasing after its mother as she dives and twists about below water. If the baby otter decided that this was not the kind of life it was seeking it would undoubtedly starve to death, for there is nobody around who will feed it rabbit and chicken in front of the drawing room fire.

The beaver, another aquatic animal, fells trees by gnawing two thirds of the way through the base of its trunk, then waiting for the tree to fall either into the river, or to make a bridge from bank to bank. Not every tree falls the right way but if enough do so, they dam the river and raise the level of the water.

The young beaver has to learn how to do this. Its instinct leads it to grind down its teeth on the tree trunk because if it failed in this they might continue to grow until they went into circles and became useless. But if it is going to gnaw through a thick trunk it may as well do so in the right place, so that the tree does not fall onto dry land.

Once again instinct may come to the rescue, prompted by a little parental tuition. The older beavers are more wary than the less experienced youngster. They like to keep their backs to the river, so they can quickly reverse into the water and stare across the river on the look out for an intruder. In this way they gnaw the side of the trunk nearer to the water. While they continue to do this they may depend upon the tree eventually

falling into the water, for it is the habit of trees to lean in its direction. But young beavers may prove obstinate, so the parental instinct comes into play and teaches them to adopt the most successful pattern of behaviour the adult beavers know of, in order to induce trees to behave as *they* should do.

As I wrote earlier, kittens have to be taught to catch and kill mice, and foxes to kill the game the father fox brings in. I owned a fox cub I took from a farm lad who had caught it in a gin, and holding it by the brush, was about to kill it with a stick. I reared this baby, not much larger than a three-months-old kitten with egg and milk fed through a doll's feeding bottle. He had never seen a struggling pullet or a young rabbit, and as he grew up he showed no desire to kill. I had a large poultry run at the time, and Bonzo spent all his nights and his spare moments (when he was not out in my car with me), fastened in the hen run, but free to gallop up and down, the chain on his collar ending in a large ring which slid along a twenty yard length of galvanized wire. His job was to keep away other foxes which would have killed the hens. Not once did Bonzo ever touch a hen, nor did any other fox while *he* was on duty. He slept in a large wooden kennel at one end of the hen run but always, I imagine, with one eye open. He had no use for carnivorous foxes and he intended them to know it.

Young birds sit in a nest with gaping beaks, through the opening of which the parent birds can recognise the identification marks of their own young ones printed on the roof of each mouth and in the pharynx. Each species has its individual pattern and the parents seldom make a mistake, excepting among certain birds such as the robins who will feed another robin's young with as much eagerness as they feed their own.

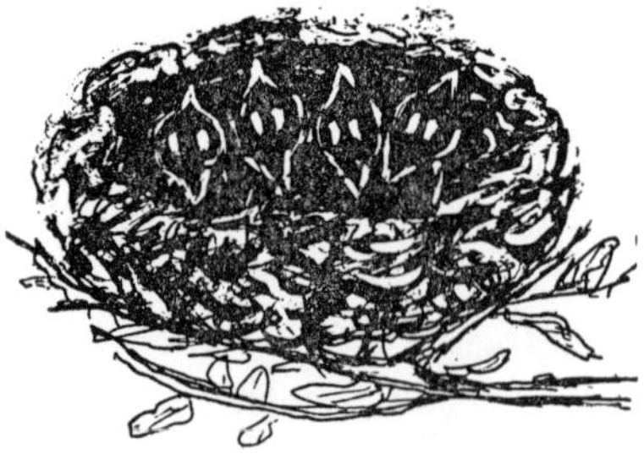

Today the mother bird may push food into the gaping beaks, but tomorrow the family may have departed. It is seldom that they can fly far, and for several days the young may be unable to rise more than a few feet from the ground, and they may then be at the mercy of every prowling cat. The co-operation existing among birds at this time will usually ensure that there are parties of birds of one kind or another on the look out, ready to divebomb the cat as soon as it makes an appearance.

The parent birds will leave the young ones to fend for themselves for an hour at a time but they are usually ready with emergency rations when they are needed.

D. W. Snow describes how pairs of blackbirds often divide up their family into two batches, one of which is fed exclusively by the father, the other by the mother. Apparently those youngsters in the care of the one partner are completely ignored by the other and if they beg from the wrong parent they do so very half-heartedly as though aware they will get nothing. The cock-bird will go on teaching the young ones, and feeding them at intervals, for the next sixteen to twenty days, during which time the henbird will be building a new nest or repairing the old one.

Young animals learn very rapidly when the tuition is punctuated by meals or tidbits and they also learn quite quickly those facts and features associated with survival. When, however, the lesson is related to more mundane matters, learning to walk on a pavement, learning to be clean in the house, and acquiring patterns of behaviour acceptable in polite society, they may appear to be quite stupid.

Puppies cannot be taught a great deal until they are from four to six months of age, or often considerably older, although they grow used to emptying themselves on a white sheet of newspaper from a very early age if this is persisted in. The sheet of newspaper can be moved gradually nearer and nearer to the door that leads into the outer world.

A great many young animals of various kinds when reared in domesticity learn to perform tricks, or special types of behaviour, in return for rewards of food. Dogs are particularly quick at associating the trick or deed successfully performed with the reward, but horses hardly ever do so. They will accept the reward eagerly enough but it does not encourage them to repeat the successful performance. A dog will usually "do it again," so long as the supply of tit bits holds out, but in all cases he will insist on receiving his reward for his last trick before he responds to the encore.

Horses are very appreciative of praise accompanied by a pat with the hand, but it is still doubtful if they know *why* they receive the praise and the pat. Certainly these do no harm, for they help to establish good relations between horse and rider, and do away with sulks and bad temper on the part of the horse, either of which is conducive to fallen jumps and loss of interest on the part of the horse in the whole performance.

Even among animals nothing succeeds like success. When success is achieved by luck rather than by judgement, the object once achieved may afford the means by which the animal learns. Once successful, not many animals can be induced to depart from the newly acquired pattern of behaviour, even when the circumstances have changed. If they are compelled to make alterations they become frustrated, and make use of substitute behaviour. In such circumstances a man will scratch his head, turn over the money in his pocket, blow his nose and light a cigarette. The dog will simply cock his leg and pass urine against the leg of the drawing room table.

To show how a new pattern of behaviour will cling, even when the need for it has disappeared, I will quote the tale of a whippet belonging to a circus, which broke both hind feet and came into hospital for treatment.

Being light of build and very lissome, the whippet discovered that it was quite easy to run about on his two fore feet, so long as the hind part of the body was elevated, with both hind feet high in the air. He practised this behaviour assiduously with great benefit to his injured hind feet which healed very quickly once the weight of the dog's body was removed from them. Soon he was quite well, and perfectly able to run about on all four feet. But the whippet, having learned his new trick, was loth to depart from it and walk again like a normal

dog. He continued through all his life to prefer his new mode of locomotion. Fortunately, being employed in a circus his behaviour pattern could be put to good account.

Even fish may learn. They soon discover when feeding time is due, and they find out as they grow into maturity that the males and females in many species have slight differences in shape and particularly in colour. When males and females are distinguishable in this way fishes will use their vision to determine their sex rather than the olfactory sense.

Although they often differ so strikingly in coloration, the males being bright and the hens dingy, only a very few birds know enough to use colours as a guide to sex. Instead, they depend upon the response to a mild display of aggression. When a cockbird makes a show of driving it off, the stranger will make a similar response, if male. If female, coy submissiveness, or a weak attempt at making a departure, will provide evidence of her sex.

Young falcons can be trained to bring down birds and deliver them to their keeper, in return for which they receive a reward of meat. Other birds, cormorants and similar birds, can be taught to catch fish. In both these performances the two birds, the falcon and the cormorant are carrying out their own instinctive behaviour. In the former the instinctive urge to consume the prey is frustrated, but by way of substitution a reward of other food is offered. In the case of the cormorant, the fish are only *partly* swallowed, a strap around the neck of the bird preventing the catch being utterly lost.

When teaching rats and other animals to learn a route through a maze, laboratory workers make use of a mild electric shock to inform the animal that it is about to take the wrong path. Sometime, instead of the electric shock the pathway to be avoided is paved in a way which makes it uncomfortable to walk on. Dogs cannot easily be taught in this way. They respond to kindness and tidbits, but will go on strike if punished, and refuse to take any further part in the proceedings.

We can only just mention the obedience classes which now constitute an important part of every dog show, to indicate the almost unbelievable things that dogs can be taught to do by training and patience.

Police dogs are equally intelligent, while the "Seeing eye" dogs need no adulation from me.

But it is not every dog that will allow itself to be brought into what one can only describe as "intimate subjection to the will of its handler." Some dogs lack the necessary intelligence or temperamentally are unsuitable subjects for training. Others may refuse to take any part in human pastimes, so different from their own conception of what constitutes acceptable behaviour.

Training depends as much upon the personality and temperament of the handler as upon that of the animal which is being trained, and this goes for all animals, from white mice to elephants.

But none of the clever things dogs do in obedience training can ever compare with the work of a pair of Welsh Collies on a mountainside. The herding instinct is very marked in these dogs and they appear to possess some intuitive ability to see into the minds of the sheep and decide how to forestall their escape plans almost before the plans are conceived. They will behave equally well in herding and driving hens, ducks, or cattle.

Space does not permit me to discuss such cleverness in detail. I have made further reference to this in *The Mind of the Dog*.

Young animals learn a great deal from play, and we will reserve the discussion of this until a later chapter.

Chapter III

ANIMAL PERCEPTION

Let us now consider some of the ways by which animals make contact with the world around them and realise the relationship they bear to their surroundings.

Perception denotes an awareness of objects, qualities or their location, engendered by stimulation of receptive media, known as sense organs.

A sense is an experience in the sphere of sensation, related to a centre in the brain, and differing from other forms of perceptivity evidenced in other brain centres.

A sensation is aroused by some stimulus external to the nervous system. It is a consciousness, excited by an external object which transmits a nervous impulse to the brain.

Among the lowest types of animals there may be no consciousness, and no sensation conveyed by an external stimulus. The stimulus, may, however, provoke an action, or series of actions, by purely reflex means.

When protoplasmic organisms, such as protozoa, are devoid of anything comparable with a nervous system, and when there is no perception of contact, there may yet be a reaction, such as may occur when a protoplasmic body approaches a food particle and merges its substance around it.

The more highly developed animals possess a number of what we term "senses," each provided with its own separate brain centre.

The majority of psychologists speak of sense organs, such as the eye, tongue, ear and nose, as *receptors.* They also favour the belief that in a number of animals the machinery may be present by which stimuli are available to the sense organs, but that the brain of the animal may be unable to interpret the impression it receives, and so no response may be made to the stimulus.

The chief senses we recognise are sight, smell, hearing, taste, touch, the kinaesthetic sense, and the sense of orientation, which relates to the position of the body in space. Psychologists are also prepared to accept the probability that there may be other senses present in animals of which at present we have little knowledge.

Each sense, in addition to possessing its own brain centre, has also its own receptor, c.f., the eye, nose, ear or tongue, while the entire surface of the body may be a receptor for the sense of touch.

The number of senses possessed by any animal may not invariably be related to the degree of intelligence present. Man, for instance, enjoys aesthetic delights presumably denied to animals, although these are to be regarded as arising only from an exaltation of perceptivity of one of the existing sense centres, one which provides an appreciation of beauty. It is believed that animals have no artistic perception, and that the song of a bird which delights the human ear may be produced quite automatically, and convey nothing to its producer. Undoubtedly it satisfies an urge, and the song may be merely a step towards the satisfaction of a need.

But while a human singer or a musician can produce music, and at the same time metaphorically step back and survey - and possibly enjoy, - the effect, it is considered that another animal has not this ability.

Whether this is fact or fancy, it is quite clear that the majority of animals possess a higher development of at least one sense than we, ourselves, do. They may have far greater power of smell, or of hearing; and while human vision is excellent in its way, there are animals far below man in the matter of intelligence, such as birds and lizards, in which acuity of vision is far greater than it is in man.

The birds which when carried away for great distances return at once to their homes when they are liberated, and the ability of birds, insects and such fragile creatures as butterflies, to migrate across wide oceans and arrive at a given part of the globe, all indicate the possession of senses of which we know nothing.

In addition to this, so many animals exhibit what we may term extra-physical, or psychic senses, or it may be a surprising development of those senses they normally possess. Such senses make animals aware of happenings of which we ourselves receive no impression. To examplify this I will quote from a paper read to the British Veterinary Association on *Abnormal Behaviour in Dogs and Cats* by Alastair N. Worden, Nutritional Research Unit, Huntingdon.

> "Belief in the popularly termed "sixth sense" of the dog is widespread, and one is always coming across almost incredulous stories - told usually in an attempt to demonstrate the dog's "intelligence." It must be conceded that we do not as yet have adequate explanations of homing behaviour or of many other surprising canine feats. I can add one *bona-fide* case, in a Scottish terrier, Sheila, who was devoted to me and who remained with my parents at East Barnet from 1940 onwards, when she was seven years old, while I was working at Cambridge. As often as I could - but sometimes at intervals of some weeks - I would return home, or call on my way to London, but my visits were usually unannounced, at different times of the day, and by any of a considerable number of local trains with which I had connected at Hatfield. Yet almost invariably Sheila would rise in pleasurable anticipation when the particular train bearing me was heard, some half mile away, approaching the local station, and thereby inform my mother of my coming. There were over 30 stopping trains a day travelling in that direction."

The above is given as one striking instance of behaviour on the part of a dog for which we can find no explanation.

Later I will quote a number of other examples.

We can only compare our own senses with those of other animals by relating the qualitative attribute, or aspect, of some experience belonging to a specific sense department, or associated with a specific sense centre in another animal, to the qualitative attribute or aspect of the same sense as we, ourselves, experience it. We can judge the modality within the animal only by the manner in which the animal reacts to it. By doing this we soon discover that the modality, or qualitative aspect, of many of the senses of animals is infinitely superior to our own, and that where one animal, c.f., man, depends largely upon one specific sense, such as vision, another animal such as the dog for example, depends far more upon smell and hearing.

In hounds, to take a further example, vision enters into the picture hardly at all, and the perception relating to tracking may depend upon a combination of scent conveyed through the nose and of taste carried back into the naso-pharynx by a handing and salivating tongue.

Then again, when man has the greatest difficulty, sometimes, in detecting a fairly distinct odour, for example when two people argue as to whether a bottle of milk, or a piece of meat has "gone sour," a male moth will detect the odour of a female from among a miscellany of scents of flowers and animals, and maybe also of motor cars and factory chimneys, and hasten to her side while she is still a mile away.

A dog can localise the direction from which a sound comes to within fifteen centimetres when its source is five metres distant. A cat is nearly as clever though not quite so discerning as the dog in this respect. Cats hear birds flying overhead but the domesticated dog of today, the product of artificial breeding, is handicapped by the size, shape and position of its ear-flaps. A cocker spaniel hears nothing from immediately above its head.

An erect-eared Alsatian hears better than the cat and far better than the spaniel.

Cats have some visual advantage over the dog at short distances, principally because they have frontal eye placement and can converge their two eyes onto an object lying ahead much more rapidly. In addition cats possess remarkable pupil control, and in bright light or under the influence of anger or fear they can convert their eyes into pinhole cameras which focus automatically. But cats, even, may not see a mouse sitting in straw until it moves. Many cats "play" with a live mouse and lose it because after the cat has freed it, the mouse will sometimes "freeze." Losing sight of it, the cat commences to search, and the mouse takes the opportunity to escape while the cat's back is turned.

Col. James Y. Baldwin* (*Our Dogs* July 29th, 1960) writing of sound and scent perception in dogs, recorded that during the war an important factory had formed a Home Guard of about forty men. In a trial the officer commanding, put all his men around the boundary, about twenty yards apart. A party of Commandos went through them in the dark as though there had been nobody there.

In the next test they put thirty-six men in reserve in a central position and sent out four patrols, each consisting of one Home Guard, one Alsatian dog, and a handler. The Home Guard man carried a Walkie-talkie. As soon as a Commando got within two hundred yards the dog scented and pointed him.

In the next trial the Commandos impregnated the air with paraffin vapour but it made no difference to the dogs.

On another occasion patrol dogs were used to detect a person hidden in one of thirty dugouts on ordinary patrol. In spite of the fact that a grass fire had broken out and filled the air with smoke, in each instance the dog picked up the scent of the hidden man and led its handler to him.

It is entirely due to the sensory perceptivity of the animal that it is able to pursue its normal individual life and survive its perils. In this way it is enabled to reproduce its kind and to help maintain the existence of its species.

For every one of these species which remains in existence, probably hundreds or thousands of others have failed to do so, and only a few of their fossils are left in memory of them. The survivors have retained the powers of perception which makes it possible to exist, and continue so to do provided always that they can still find suitable environment, and that no other species in the course of its evolutionary progress develops modifications with which they are no longer able to contend.

Two of the factors against which animals are more or less powerless are accident and certain types of disease.

* Col. Baldwin was the first to employ Alsatians for army and police work and to train them for their duties.

Accident must include world upheaval, whether its origin be internal or external, and from the viewpoint of the animal the type of world upheaval which has aided most in its extermination is the advent of mankind into its territories.

When an epidemic strikes a community of any one species, it is seldom that all its members succumb. Frequently enough escape or recover to re-establish its numbers after a variable period.

There may seem to be little connection between an ability to withstand disease and the operation of the senses, but quite frequently the senses play a greater part than is at first apparent. The rabbit, for instance, which avoided being bitten by rabbit fleas, and possessed sufficient perceptiveness to avoid flea-infested burrows, had a far better chance of avoiding myxomatosis, and producing progeny with similar characteristics.

Some animals, fishes for example, possess sense organs somewhat different from those common to most animals.

FISHES

A great many fishes (but none of the herring family) exhibit quite plainly a development known as *"the lateral line."*

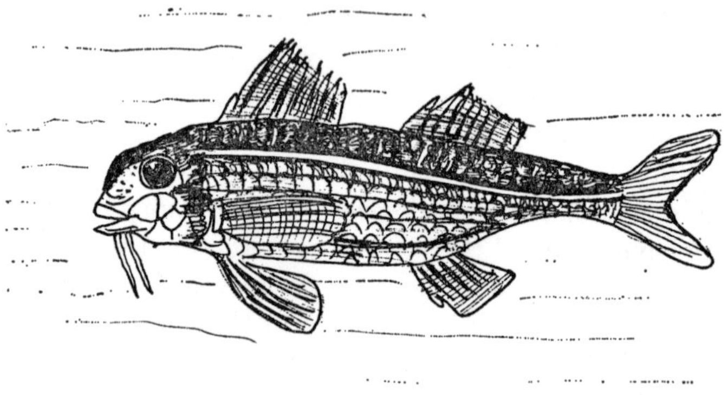

You can see this very clearly on a cod or a whiting. It commences, not so noticeably, around the eyes and over the head. Then it extends quite visibly down the length of the body on either side.

The surface of the line is perforated by small apertures lying between the scales, and the "line" is in reality a tube within the skin, containing some baffle plates and a quantity of a thin, jelly-like fluid, in which a number of sense-cells and nerve fibres may be demonstrated.

These cells convey to the mind of the fish some impression concerning how far it lies below the surface, the relationship of its body to rocks and weed, or to other fishes, the strength of the currents and the amount of movement in the water itself, as well as some indication of its temperature. It may also co-operate with the gills in estimating the concentration of salt in sea-water.

Every female fish when about to select her spawning ground has to make sure that the degree of salinity will be suitable for the hatching of her ova, and that it will not cause them to shrink, or to fill up and burst. In some cases the male will have provisionally decided where the eggs are to be laid (as in the sticklebacks) but as soon as the female arrives on the scene she may realise instantly that he has hit on an unsuitable location, and will whisk him off to another region immediately.

The lateral line is sufficiently sensitive to enable the fish to swim safely in complete darkness, avoiding all solid objects. It is said that a fish provided with a lateral line can recognise the "flip" of the fin of a larger fish which might prove to be an enemy, in its vicinity.

Fishes are all the more dependent upon this extra sense receptor because according to our own standards of hearing fish are comparatively deaf. They have no external ears and no auditory canal leading into the middle ear such as most other animals possess, although they are provided with an internal ear of very similar type to that present in most of the mammals.

The ears of fishes contain small "ear-stones," more generally known as "*otoliths*." These are white pebbles which increase in size as the fish ages. They are contained within small sacs, and the organ thus produced takes the place of the middle ear of mammals, and endows the fish with the power to balance the body in water, of knowing which way up it is (orientation in water), and it also gives some indication of whether the fish is swimming upwards towards the surface, or downwards towards the bottom.

We used in our early days to be regaled with stories of monasteries, in which there were fish pools containing monster carp. On the ringing of a bell these fishes used to swim to the banks to be fed by hand.

It is quite true that fishes see quite a lot of what goes on ashore, according to the amount of light, the area and depth of the pool or river and the contour of the immediate surroundings. Fish are also very aware of ground vibrations arising on shore and transmitted to the water. It is more likely that the monastery carp became aware of the preparations by the stamping of feet, and if the bell were capable of a resounding clang it is possible that it might set up minute vibrations in the water itself, which would be communicated through the middle line but not necessarily interpreted by the fish as an auditory stimulus.

All the carp family possess the lateral line and may well have an ability to "feel" vibrations rather than hear them in the way we do.

The above indicates very clearly that all the senses of animals are not identical either as regards receptors or the interpretation of stimuli, but that, nevertheless, the end result, transformed into activity, may show a good deal of similarity.

There is one other feature in the carp family that may perhaps have some further bearing upon the behaviour of monastery specimens.

In common with quite a lot of other fishes, which live in fresh water, carp belong to the *Ostariophysi,* which include about five thousand species. In this group the air bladder communicates with the middle ear by means of a set of small bones not unlike those present in the internal ear of mammals. Whether these render the members of this family any less deaf is still uncertain but it is possible they may assist in picking up vibrations from land or water.

Fish are presumed to manage without conversation, so it may well be, as I remarked above, that vibrations convey as much, possibly more, information than sound waves would do.

Perhaps I should qualify the statement regarding piscine conversation since Rachel Carson in "The Sea Around Us" assures us that the deep sea, which we had thought of as silent, really reverberates with strange sounds, with barkings and boomings and weird whistling noises, comparable with Euston Station in a blackout.

A few animals possess anatomical structures the use of which is unknown. They may represent remnants, vestiges of some organ present in remote ancestors, or they may be special sense organs. Very little is known about them.

Some of the lizard family, and the lampreys, parasitic fish which attach themselves to other fish and slowly eat their living bodies, carry vestiges of a third eye, which in the lamprey admits light into a small elementary retina. It is unlikely it produces an image or a mental impression of one, but it at least gives information as to the degree of illumination overhead.

The *pineal gland* is a small structure present in the brains of mammals, while in fishes it is attached to a stalk. It probably represents the remains of a third eye although it seems to serve no purpose at present. At one time men regarded it as the seat of the "soul."

In calm water, fish which live in rivers, lakes and ponds are able to see a good deal of what goes on ashore, not directly, but by reflection onto the surface of the water. When the water is clear and the illumination is good, the fish looking upwards sees a circle of light - a window - and in this is reflected a picture of the surroundings, which includes everything visible between the centre of the window and the horizon. Objects nearest to the centre, such as the tip of a fishing rod, the end of an oar or paddle, or the overhanging branch of a tree, will appear quite big to the fish, while the fisherman sitting on the bank, even if only a few yards distant, will look like a tiny doll.

The deeper the fish lies in the water the larger the circular window will appear, although if the light is poor and the water cloudy, the amount of detail seen may be negligible. The deeper it lies, too, the easier it will be for the fish to see any movement on dry land. When the fish approaches the surface the window becomes very small but the picture will be clearer. Around the circumference of the window of light, the fish will see a second concentric circle and in this will be pictured the bottom of the river or pool, if not too deep, together with objects resting upon it.

If the fish looks up at the circle of light it may see through it any fly settling on the surface, and it may possibly see beyond this, since the Archer fish is able to spit a column of water at a fly twelve or eighteen inches above the water and bring it down. When we count the number of refractive surfaces light must pass through on its passage from the fly to the retina of the fish, it seems almost impossible that the archer fish can make such a hit without an enormous amount of mathematical calculation!

Smell in Fishes

With the exception of the lampreys (which are not true fishes), the nostrils of fishes do not communicate with the pharynx but lead into a double-walled sac lined by a membrane containing many nerve fibres, which join up with the olfactory nerve. Most fishes have a good sense of smell for in many cases their prospect of living will depend upon it.

In sharks and dog fishes, bass and wrasses, all partial to decomposing material as food, the sense of smell is especially keen. Sharks can smell blood in the water, or a dead body even when fresh, over a distance of up to two hundred yards. If, experimentally, their nostrils are plugged to eliminate all sense of smell, sharks will starve to death in the midst of plenty. Vision by itself is insufficient.

Taste in Fishes

It is unlikely that many fishes possess a very highly developed taste sense although some habitually take food into their mouths, forcibly eject it then repeat the process, often several times before they decide to swallow it or finally reject it. When they make up their minds to consume it they bolt it whole with no attempt to chew it.

Many varieties have barbels or wattles in their lip region which may be able to taste food. These barbels are also used as feelers to detect the nature of objects by the sense of touch. Other fish have taste buds, or nerve centres, in the roof of the mouth and on the lips.

Quite a number of fishes swallow their food, even hard crabs and shell fish whole, while others such as mullet and the Gillaroo trout have thick-walled stomachs which crush the shells of snails and small crustaceans with ease. Members of the carp family carry their teeth in their throats.

Movement and Vision, in Fishes, and Animals Living in Water

Like the majority of animals fishes are greatly attracted by anything which moves, and a still object even more attractive in the food sense, may be completely ignored.

Anglers take advantage of this when spinning. Not only does their bait move through the water but it also moves upon itself. A fish may reject a dead worm, or motionless maggots, however cunningly coloured, but will accept an artificial worm if an appearance of movement can be imparted to it. If a piece of fine, insulated wire, red in colour is bent in the shape of the letter "N" and towed on an invisible nylon thread carrying a small hook or allowed to travel in a current, the bait will catch fish because the bending of the wire gives the fish the impression that it is a living worm.

An octopus is accustomed to catching crabs on a sandy sea-bottom. Should the octopus be lying within a rock crevice and a crab is dangled in front of it, suspended in water, the octopus will ignore it since the visual pattern is not the one the octopus connects with food and feeding. It does not act as the approved stimulus - it does not ring the bell.

In a similar way a pike, unless particularly hungry, will pay little attention to a dead roach suspended on a hook, especially if it is dangled at one side of the pike. If the bait lies straight ahead and shows movement, the pike is likely to make a rush at it, and even when it is made up, perhaps of plastic or metal, with little of the appearance of a fish, and is jerked away, the pike will almost certainly seize it.

As I will explain later when dealing with the Gestalt theory of visual perception, movement may attract attention when nothing else in the entire visual pattern is making any mental impression.

Movement within the visual field may act as an immediate stimulus and this may be translated into equally rapid activity, even when the impression is based upon entirely false premises.

Fish are very sensitive to sudden changes in water temperature and tropical fish kept in aquaria may die overnight if their water becomes chilled just as much as if it becomes over-warm.

There are scientists who doubt whether fishes really see in the same way that we do, by recognising an image and considering it in the conscious mind. In some ways the brain of the fish is very undeveloped, the lobes concerned with sight especially so, in many varieties. The question the scientists ask is: Does the visual image set the appropriate response into operation quite automatically, without the fish receiving a visual mental impression in the way that we do?

However, few anglers would accept such a theory, but it may cheer them to read that the automatic response of a fish to a yellow maggot may still be quite different to the response made by the same fish to one dyed red, even if the fish "sees" neither.

SNAKES

I have already mentioned the ability of the mosquito to detect minute changes in temperature and this applies equally to most of the blood-sucking parasites.

We must now pay a little further attention to the rattlesnakes, copperheads and mocassin snakes, the three of which are usually classed as "pit vipers," on account of an unusual sense receptor located in their heads. In each of these snakes a pit is visible between the nostril and each eye. This spot acts as a thermometer and by registering a minute increase in heat from an animal body, it informs the snake where to strike. On this account the pit vipers are far more accurate and much more deadly than other snakes which rely on vision. The pit viper can detect a variation of $\pm$ .2°C. in the temperature of a body or emanating from it. If a slightly warmed and an unwarmed object are swung alternately in front of the snake it will strike the warmer one each time and ignore the cooler. This happens even when the snake is blindfolded but if the pits in the head are plugged with collodion or wax, the snake may appear confused and refrain from striking.

In a fight between a cobra and a mongoose the cobra may strike again and again without success, since the mongoose crosses the line of vision of the cobra so rapidly. The snake has never learned to aim in front of the mongoose, or to get in line and strike parallel with the ground surface.

Snakes having one eye on either side of the head can, with one or two exceptions, see an object with only one eye at a time. They are compelled to wriggle their heads into various positions in order to get an all-round view. Accordingly, they fail to make use of binocular vision, with the result that their aim is bad.

Hold up your index finger eighteen inches in front of your face. Close one eye and note the position of your finger against the background. Now open and close each eye alternately and see how the position of your finger against the background varies. This will give you some idea of the difficulty the snake experiences when compelled to use its eyes separately. This is why cobras miss so often, but a rattlesnake "aims blind," being guided purely by heat impressions received through its face pits.

Nevertheless, snakes possess quite good vision, and ability to focus accurately. Those which sleep by day, and move about by night, depend wholly upon smell and taste.

The nocturnal snakes follow their own trail away from their homes and back, and they do this by scent and by impressions received through their flickering forked tongues which convey sensation to the brain through a special sense receptor known as Jacobson's organ. This is present in most reptiles and in some other animals. It is present in the human embryo but becomes absorbed during its development and has completely disappeared at the time of its birth.

Snakes which travel at night, "blind," follow the same path on each trip, so that if a traveller pitches his camp on the route used by one of these snakes it may travel across his sleeping body on the way out and return across it on the way back. If the snake happens to be a female, a male may trail it by smell, and in this case the traveller may sleep particularly badly.

This combination of taste and smell is not confined to the snakes but is shared by the hounds, which pick up the scent they follow both by smell and by the taste of the fox or stag, or man, in the moisture of the earth or grass, while their tongues are out and trailing saliva. The combination of smell and taste takes place in the naso-pharyngeal region.

Neither do *we* need to be reminded that our pleasure is doubled when a meal not only tastes but smells "good."

Since most snakes have no true ears that one can see, and also because they hear badly, they depend a great deal upon ground vibrations conveyed to the hearing apparatus through the bones of the chin, which they rest upon the ground.

Indian snake charmers make great play with their pipes or reeds. The snakes have been kept for some time in complete darkness, and when the lid of their box is raised they lift their heads while half-blinded by the light, and concentrate upon the hand movement of the player. It is customary to close the lid of the box before the snake recovers its vision sufficiently to make a move towards freedom.

LIZARDS

Lizards also make use of both tongue and nostrils to smell their prey, and also to find their way about their own particular territory. Without their sense of smell and Jacobson's organ they would be greatly handicapped. They are provided with a cone-shaped area in the retina. This greatly resembles the same area in man and, accordingly, the lizards possess marked acuity of vision. Like the snakes they are dependent entirely upon monocular vision, but they have excellent peripheral vision, too, which means they can see in all directions and not only centrally. The peripheral portion of the retina is especially good at detecting movement.

INSECTS

Vision

In company with other animals such as the lobsters, crabs, shrimps and pawns, a great many of the insects, particularly those which fly, possess compound eyes, often together with some simple eyes, frequently three in number.

A compound eye is made up of a number of separate columnar units, each of which contains a cornea, the transparent window of the eye. Each window is hexagonal in shape, and as adjoining windows are united by their margins or frames, they combine to form a large visual surface. Each columnar unit (*ommatidium*) contains a transparent cone which acts as a lens, and also a structure which takes the place of the retina in a simple eye, connected by nerve fibres to the sight centre.

Inside each cone is a quantity of black pigment (melanin) which moves about to occupy various positions within the interior of the cone, so that in a bright light it cuts out some of the rays in much the same way that the iris does in the eyes of the higher animals.

The first thing that strikes the observer when looking at the eyes of wasps, bees, dragonflies, butterflies and the like, is their enormous size in comparison with that of the head and body. Not only do they occupy the greater part of the head but they bulge outward, so that the insect can see in all directions at one time.

The number of facets or windows in the eye of a dragonfly has been counted as 12,544, in certain beetles over 25,000, in butterflies 17,355, in the housefly 4,000 and in ants, 50.

The simple eyes are of no great value for seeing but they admit light and assist the action of the compound eyes.

In spiders, which are not insects but members of the Arachnidae, there are only simple eyes. These are usually carried on top of the head and there may be as many as fourteen, or as few as four, according to the variety of spider.

It is very unlikely that the compound eyes of insects and crustaceans see a large collection of separate pictures but rather that each unit sees a separate portion of the entire picture. The smaller pictures will probably combine within the visual centre to be interpreted as one composite whole. It must not be supposed that eyes of this kind provide vision over any great distance, but they suffice for the needs of their owner, and are particularly good at detecting movement, and when this is observed by its eyes the flying insect automatically takes off.

Many flies, c.f., houseflies, have but little auditory acuity. One can shout at a housefly on the tablecloth without result, but the least movement in its direction, or the casting of a shadow across it, will cause it to take off in flight.

Honey bees and some insects, notably water fleas, see ultraviolet rays which are invisible to man and most other animals.

When bees are attracted by red flowers it is not on account of redness, but because they contain other colours also. The bee sees no red but is attracted by the other colours which become visible to its eyes.

Hearing in Insects

Much of the sound in the insect world is related to courtship and mating. We have all heard the rasping squeaks produced by grasshoppers, and possibly by crickets.

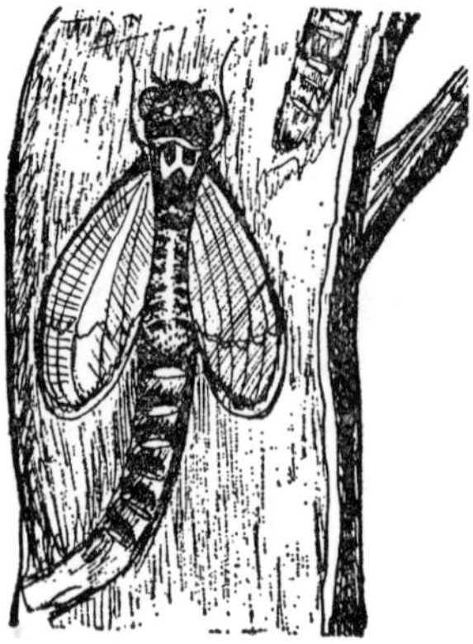

In tropical forests one of the most constant and continuous noises one hears is caused by insects calling their mates, and of these the songs of the cicadas stand supreme. Not on account of their beauty it must be admitted, for the vocal efforts of the cicada have been compared with a good many other noises, and the most apt comparison was probably one which declared that the sounds ranged in character from the noise made by a duck laughing, to that made by a hack saw working its way through a loosely-held sheet of corrugated iron.

In spite of this, the cicadas know precisely the meaning of each change of tune. These variations are affected by emitting the song in short pulses of sound in rapid succession, with frequent variations in amplitude and time. Fortunately, only the males sing, a fact which led Xenarchus, a poet of Rhodes, evidently a keen observer, to say "Happy Cicadas! Thy females are deprived of voice!"

The song of a cicada can be heard at a distance of half-a-mile, so when twenty thousand of them, all in a mood for courtship, congregate in the same small belt of trees, their musical effort can be better imagined than described.

A curious thing about cicada song is that it may be sufficiently deafening to keep one awake; then, quite suddenly, it ceases for a few seconds. Why this happens is not clear but the signal must be received by every cicada at the same moment and instantly obeyed. One is reminded of the "hawk alarm call," universal among the birds, which, uttered by one individual, causes every variety of bird to "freeze" into silence and immobility.

The stridulation of grasshoppers and crickets is produced by rubbing the surfaces of their thighs against some serrated part of the body. This produces a chirping note which announces the insect's whereabouts and the fact that the male grasshopper, at least, feels in a mood for matrimony. The crickets are remarkable for the ardour exhibited in their lovemaking, which bears a marked resemblance to that prevalent in higher animals.

If a single chirp of a male cricket comes over the air, more particularly if it be derived from a gramophone record, every female cricket in the area will be attracted towards the instrument. When the sound is introduced through the telephone, and the receiver is left lying on the floor, the local female crickets will make one concerted effort to reach it first and climb into the mouthpiece. They will persist in this unreasonable behaviour even when they should be aware that scores of perfectly eligible male crickets are loafing around. These appear to be quite invisible to the excited females, so long as the voice comes over the 'phone, and in spite of the fact that the cricket which produced the song has been dead for many years, the charm of his voice still lingers.

Perhaps, after all, it is a wise grasshopper, or cricket, who does his courting by 'phone, for it is quite likely that the passionate female will commence making a meal of her bridegroom during the act of mating.

In the case of some insects, notably the Mantides, it is by no means unusual to see a female mantis, who may be several inches in length, calmly sitting on the half-eaten corpses of her first four suitors, in the act of mating with the fifth, from whom she is daintily severing the head.

Some of the lesser bugs have nicely modulated voices, audible to our own ears at only a few inches distance. The sense receptors of insects which receive auditory impressions may be in their legs, particularly at those portions which bend, such as the knees, or they may be associated with any of the joints of the body itself. "Ears" of this kind situated in the limbs are particularly useful for picking up ground vibrations.

Some of these *chordotonal organs* are partnered by an air sac provided with a tympanum or ear drum, but in a great many insects, of which the mosquito is the best example, both hearing (reception) and sound transmission are carried out by Morse code. The aerials are represented by the antennae which are covered with fine bristles after the manner of a bottle-brush. Transmission is effected through the bristles of the female and received by those of the male. There is no record of whether the male is able to make a reply, and in any case there would be little time, as the moment the message comes through the male turns himself about until he is getting the maximum reception, then he tunes in on the wavelength and is on his way; and in spite of all jamming, he keeps on the right course until he meets the lady who has despatched the call.

Although many of the insects employ only a limited waveband, a number of others make use of supersonic sound, which though inaudible to our ear is heard quite well also by the animals which seek these insects for food. It is believed that the hawks and falcons, as well perhaps as the buzzard which flies high, and the owls which fly low, are led to their victims as much by sound as by vision. It is likely that cats and dogs hear supersonic squeakings of mice and rats. Certainly, they are quick to answer the supersonic whistle, and as it is well-known that many small animals employ supersonic sounds as their own method of communication, it is likely that the dog and cat hear them when we cannot do so.

Other Means of Communication

Many of the moths which have a very limited ability to produce sounds make contact with their mates by means of odour. The female moth, when receptive, protrudes the odour-emitting organ from the hinder end of her body. The perfume wafted in a gentle and favourable breeze is recognised by a male at anything up to a mile distant. As soon as this happens he flies to the place where the female is resting. He is unable to recognise her as another moth until within a foot of her body. Even then he may be uncertain since he will mate equally well with a pill-box or a scrap of paper carrying the odour of the female.

Ants, many of which are blind, appear to communicate mainly by touch, by stroking the bodies of their companions with their antennae and limbs.

The termites are blind and live in total darkness inside their homes. They build enormous nests which may stand ten feet above the ground and are often sufficiently strong to carry the weight of a buffalo, which may use the anthill as a vantage point for obtaining a better view of the surrounding country.

One West African termite, known as the "snouted harvester," carries a gland in its head which secretes an irritant fluid. It squirts this through a hole in its snout and lays out its enemies among the other ants. How blind termites distinguish friends from foes is not easy to understand.

Movement

It is interesting to speculate whether flying insects are conscious of their flight through space since the actual wing movements which occasion the flight are entirely automatic, although they can be brought under control, as for example when a wasp wishes to settle on a banana. What we would like to know is whether the wasp imagines itself to be motionless with the scenery moving past it, or whether it aims at some definite object and flies towards it quite consciously with the intention of settling upon it. In the first place one would need to know the range of a wasp's vision. Possibly it is at the most two or three feet, and quite likely less than that, so unless the wasp flies close to a wall or shelves, or in a narrow passage it will see no objects passing it during its flight. It is likely that the wasp that flits from one teaparty to another on the open beach, or chases after a little girl who is running about while at the same time she takes bites at a banana, is hunting entirely by smell. A wasp can certainly pick up the scent of an orange or banana, or of a fruit jelly, from a distance of at least fifty yards and fly straight towards it. When it comes to within three feet of the origin of the scent, this will become so overpowering that the wasp may not be able to locate its exact origin, and then at short range sight may come to its aid.

One would imagine that a large number of facets in a compound eye would produce some blurring of the composite image but it is likely each unit receives a separate impression during flight, embodying only a fraction of the complete image, and that a rapid succession of pictures passing over the retina-like part of each unit may be built up into a picture in the same way as a rapid succession of pictures on a cinema screen can be translated into a moving picture.

Not only do insects possess an excellent sense of smell but they also have marked taste discrimination. Butterflies can detect sugar in a dilution, two thousand times weaker than the human tongue can detect it. Houseflies will take sugar with avidity but refuse saccharine.

SHRIMPS

The compound eyes of shrimps enable them to see through the water of a rock pool and recognise movements around its edges. If you kneel down at the side of the pool and hold your hand above its surface with one finger raised, all the shrimps will depart. If you remain motionless for one or two minutes they will return. Then, keeping quite still otherwise, lower your finger. The shrimps will recognise the movement at a distance of eighteen inches from the surface of the water and will again dash for safety.

BIRDS

When we survey the animal kingdom in search of the creatures which enjoy the greatest advantages in life, coupled with the fewest disadvantages, I think we should award the prize to the birds. They are blessed with wonderful vision, excellent hearing, and a useful sense of smell, greatly enhanced in those birds which have a taste for carrion. It is possible of course that if we could change from a human being into a bird we might miss some of our present epicurean delights, since a diet of seeds might not appeal to all of us. But as one man's meat - and presumably one animal's - is another's poison, it is equally likely that my canary might be unwilling to exchange his mixed seeds for a plate of steak and chips. Whether the bird has a sense of taste which can find pleasure and variety in dry seeds is difficult to say, but as my own bird throws nine-tenths of his seed on the floor in his search for delectable varieties, I must presume that he finds these very much to his liking. If I change his brand of mixed seed he goes on hunger strike until I return to his customary packet although so far as I can see there is no material difference.

Poultry, I am always told, should never be given whole maize as a staple diet, but my own birds always picked out the maize first, the barley or wheat second, and the oats last. Once again, they showed a taste preference, unless colour means more to the domestic hen than flavour. Certainly, the grains of maize stood out from the rest in brightness, and it is known that bright colours stimulate appetite, while dull, drab shades depress it. The confectioners will agree with this statement.

One of the greatest advantages which birds possess is the gift of flight, a rapid and seemingly not too fatiguing means of travelling a long distance, by the shortest route, in the least time. A bird sees the whole of the countryside or the layout of a city from an original angle and no living creature has better opportunity of gaining knowledge of the land in which it lives. A great many birds when once the difficulties attendant upon migration have been overcome, enjoy the best of two worlds. It may, or may not, be considered an advantage, also, to wear a suit of feathers, cool in summer and air-conditioned in winter, and waterproofed when the environment demands this kind of protection. Feathers are an excellent means of maintaining an equable body temperature - but there are times, I feel sure, when we ourselves would choose a tailor in preference to a moult.

Bird Vision

In *Animal Vision* (Jenkins (Herbert) Ltd.) I have described various types of eyes found in birds and the means by which they focus their eyes. Birds have a special type of retina and often two *foveae,* centres of marked acuity. As a result of this, a falcon hovering in the sky can see a moving mouse a thousand feet beneath, and a female peregrine observed by the zoologist Dr. Bastian Schmid, distinguished pigeons feeding in a cornfield from a height of three thousand five hundred feet, and swooped down upon them. The maximum height at which falcons have been known to see objects on the ground is five thousand five hundred feet.

Even the smaller kinds of birds possess remarkably good vision. David Lack, in *The Life of the Robin* (A Pelican Book), states that the cock robin is able to recognise his mate at a distance of at least thirty yards. As all robins look alike to us, both sexes being red-breasted, this seems to be a remarkable feat of vision.

It is even more astonishing that a robin has been seen to

drive away every trespassing robin immediately, yet never make the mistake of shooing away his own mate.

It is believed that birds recognise their own kind, and one another, by facial and physical characteristics, but one can never be quite certain whether hearing may not play a part, without our being aware of it.

The voice of the bird, a chaffinch for example, is made up of tones, some of which fall within our range of hearing and others which are outside it. Thus, when we listen to the "trill" of a songbird we must remember that it is built up of an alternation of high and low tones. The human ear distinguishes the higher tone but may be quite unconscious of the lower. There are two considerations in this case:-

The first is that to birds, and to many mammals provided with hearing apparatus far more efficient and discriminating than our own, especially as regards tone, inability to hear the lower part of the trill would rob the song of its entire meaning.

The second is that if birds are able to produce a song, only one half of which registers in the human ear, it would appear that birds might well be capable of communicating with one another using tones quite outside our own tone register. When we imagine that two birds recognise each other on appearance alone, it may also be possible that they may be making use of sounds quite audible to themselves, which we are quite unable to hear.

In a great many varieties of birds (which possess colour vision) there seems to be complete inability on the part of the gorgeously-apparelled cockbirds to make any practical use of the fact that the hens usually wear dingy colours. One would think they would recognise them as females at a glance, and that they would know that every smartly-clad bird was a male - quite a contrast to the respective appearance of the sexes in the higher circles of the animal world.

So instead of making use of what to us seems so obvious an indication (when once we have been told, as birds never are, of course), the male bird, ensconced in his territory, will set eyes upon another bird of the same species and follow it up with a display of male aggression. If the stranger happens to be male also, it may return the compliment by a similar *show*, or, it may beat a carefully-casual retreat. But if the intruder is a hen, instead of a display of aggression she will simper and look coy, possibly inviting further acquaintance.

Monocular and Binocular Vision in Birds

The majority of birds, especially the small song birds, see with only one eye at a time, with little binocular forward vision. There is usually a blind spot immediately ahead of them, since the eyes are placed well back and the prominent forehead prevents both eyes seeing an object in front of the beak. This is why pigeons bob their heads from side to side so that they can see the food they intend to pick up and swallow. A good many of the insectivorous birds which catch flies in the air, such as the swallows, swifts and flycatchers, have rather better binocular forward vision, while the birds of prey see well, and can make use of binocular vision or monocular lateral vision at will. The wading birds have eyes set so far back that they see more behind than in front of their bodies, and this also enables them to watch out for the hawk overhead while their long beaks are buried in the mud or sand.

The owl is an example of a bird with frontal vision almost parallel as in man, with practically no eye movement to enable it to see on either side. This is amply compensated for by an extremely flexible neck and although the owl sits hunched up with its head so tucked into its shoulders that it appears to have no neck at all, it can move its head through a range of 240°. It can sit with its body facing forward and its head facing backwards, and it can do this in two positions. In one the head is right way up - in the other it is completely upside down.

Night and Day Birds

We must divide birds into the durinal which fly by day, and the nocturnal which fly by night and hunt during the hours of twilight and more or less through the hours of darkness. The diurnal birds comprise the greater part of the bird population and have the disadvantage of being practically blind from an hour after sunset until sunrise.

Most of the nocturnal birds see reasonably well in the dark and require only from a tenth to a hundredth of the illumination which human eyes must have if they are to see anything at all. It is nowadays denied that owls can detect the living bodies of mice by the infra-red rays they emit. It is more probable they hear their supersonic squeaking.

The eyes of hunting birds are placed more frontally than those of the hunted, but in the latter the eyes bulge further out of their sockets, so that they see behind them as well as all around their bodies. This permits more light to enter the eye, and although this may not improve definition it enables the bird to use its peripheral vision, and this is particularly useful in providing warning of anything moving in the birds' vicinity. Whether the bird be the predator or the potential victim, it is more important to see the moving bird than to see one motionless, and possibly hiding itself away.

Ears

Mammals have one feature which birds do not possess - earflaps.

This does not appear to act as any great handicap since all birds appear to hear well. The domestic hen is particularly good at locating the distance from which sounds originate and her chickens are responsive to every tone of their mother's voice. In birds the ears are situated behind eye level on either side of the head and when compared with those of mammals which carry their ears high on the head protected by flaps, they seem equally capable, since the side of the face takes the place of the ear flap. Birds have such flexible necks that they can turn their heads and ears very rapidly in every direction.

Man is about the only mammal lacking ability to waggle his earflaps and this is due to the fact that his ears, like those of the hen, are carried laterally on his head, and his fixed earflaps enable him to distinguish between sounds coming from before and behind. Although his neck is short it is sufficiently flexible (in youth, anyway) to locate the origin of sounds reasonably well.

Animals, such as dogs, with moveable flaps, "talk" with their ears and this ability is marked in the horse and dog. Hares and rabbits have the greatest ear movement and the donkey follows them closely. Cats, rats and mice appear to have the least movement.

It is by watching ear movement that a human being can obtain a good idea of what is going on in the animal's mind. Perhaps this is another reason why our own ears no longer waggle.

Eye Protection

The third eyelid is well developed in birds, and in some, particularly in the diving birds which swim under water, there is frequently a small transparent window through which the bird can watch the fish it is chasing without experiencing irritation from its passage through what is often none-too-clean water. Cormorants are able to follow up their prey in this way at a distance of ten feet, which is their optimal point of underwater vision.

The cat family make good use of their well-developed third eyelids, as a shield against sunlight, while the polar bear uses his semi-transparent pair as a protection against snow blindness. The bear also sits shielding its eyes with its hands and often looks out between its fingers to cut out glare.

Fishes have no eyelids at all, with the possible exception of the mullet and some of the shark family. They are often provided with "spectacles," a horny transparent layer lying over the cornea, which keeps sand and debris from entering the eyes. Few fish see further than fifteen feet unless the underwater conditions are particularly favourable.

Eye Placement

The life of every animal is influenced very greatly by the position which its eyes occupy in its head.

In man and the great apes, they are situated frontally, so that man is said to have parallel vision. Nearest to man come the domestic cat and the owl, closely followed by some of the short faced dogs. Usually those with frontally-placed eyes are the predatory animals - the hunters - and they include the birds of prey and the cannibal fishes.

In most of the smaller animals the eyes are placed almost laterally. In rabbits and hares they are carried so much at the side of the head that these animals find it almost impossible to look straight ahead with both eyes at one time. They are able, however, to lay back the ears, raise the chin and watch every move of a pursuer, but may then have difficulty in looking ahead. Coursed hares sometimes run among the spectators and have been known to gallop over a cliff top, sometimes with the greyhounds following them.

Eye Placement in the Dog

The dog requires individual consideration, since among the six hundred breeds of domesticated dog there is quite as much variation in eye placement as might exist between two animals as diverse as an owl and a squirrel.

During the evolution of the domestic dog from its original ancestor, the wolf, two types of head made their appearance.

(a) The *dolichocephalic;* long in skull and jaw, after the style of the wolf;

(b) The *brachycephalic;* round-headed and short-faced, with very short jaws.

In (a) the eyes are carried a little obliquely, i.e. neither frontally nor entirely laterally.

In (b) the eyes are carried frontally, and in some breeds as much so as in the owl.

Examples of the dolichocephalic dogs are the Alsatian, all the running dogs from Afghan to Whippet, and most of the terriers. The brachycephalic breeds include such dogs as the Pekingese, pug, bulldog and Griffon.

In dogs of the long-headed type (a) the eyes can be directed forward by a slight effort, so that both eyes see an object ahead at the same time. But, normally, the eyes are set at an angle of about thirty degrees with the middle line of the head, while in the short-headed group (b) the direction of the eyes is nearly parallel with the middle line, so that they look forward the whole time, just as do the eyes of the cat.

The short-headed and short-muzzled dogs which appeared, probably millions of years ago, as a "sport" from the prevailing long-headed dogs, had underdeveloped jaws, often also undershot, and in addition they had short legs. They would have been quite unable to kill their game even if they had been able to catch it. The only way such animals could exist was by becoming scavengers and they probably hung around human settlements hoping someone would attempt to drive them away by throwing a bone at them. In this way they eventually made friends with mankind and became domesticated. There is a further theory that man first caught these dogs as puppies, and reared them with an eye to the pot, just as in earlier days the Aztec monks used the modern Chihuahua first as a bed-warmer and later, as a stew.

One of the strange things concerning these short-faced dogs is that they possess the frontal vision characteristic of the hunter, to a far greater extent than the dogs such as greyhounds and Salukis which are the true hunters.

Not only is the life of an animal influenced by the placement of the eyes within the head but it is almost equally influenced by the eye level, as compared with the ground the animal stands on.

Some dogs are small enough to get into a quart jug, so their eyes are only a few inches above the ground and their range of vision is correspondingly short. These show far greater interest in life when they are carried and can look around.

In a Great Dane the eyes may be three feet from the ground.

Most of the brachycephalic dogs have large prominent eyes, often myopic (shortsighted) while the greyhounds and running breeds are almost invariably longsighted (hypermetropic). A Saluki can see a gazelle moving about, a mile away.

Foxes and badgers favour hilltops from which they can survey the country around themselves.

In snakes the eyes are carried close to the ground so they see for only a short distance. This is one reason why snakes climb into trees. The East Indian Tree Snake, has frontally-placed eyes with a groove from each eye down to the nostril along which it takes aim. It is the finest marksman among all the snakes.

BATS

Bats have very poor day vision but see fairly well by night, although they depend for safety during flight upon their ultra-sonic voices and radar-like perceptivity.

The hearing of the bat is hypersensitive. It can recognise the faintest variation in the intensity of the note it utters when it is echoed back from any obstacle. This may be insignificant in size and substance and it is said that a bat can avoid flying into any object no thicker than a human hair, and can detect an object the size of a shilling while still in flight and thirty feet from it.

All this depends upon normal hearing for if the ear passages of a bat become thickened, or infested by parasites, the bat is no longer able to avoid flying into objects, nor can it keep its true course. This may be one reason why a bat occasionally flies into a room at night through an open window, and ends up, if country gossip is to be believed, "tangled in a woman's hair" - something which no normal bat would ever permit itself to be.

The ultra-sonic emissions which issue from the bat can be varied in accordance with its needs.

Professor R. J. Pumphrey, Derby Professor of Zoology, Liverpool University, has shown that in the leaf-nosed or horse-shoe bats the nostrils are about half-a-wavelength apart. The sounds emerge through the nostrils and not through the mouth, which is held open to receive flies. The shape of the nostrils together with the extraordinary shape of the facial structures, causes the sound to be emitted in a fan-shaped beam directed straight forward. In contrast with other bats the pulses are long, so that the leading edge may be reflected off an object thirty feet away and return to the ears of the bat before the trailing edge has left the nostrils.

When the bat is at rest there may be ten emissions per minute of ultra-sonic sound, running to twenty when the bat sets off in flight. This rate is fairly constant until the bat is approaching an obstacle when the rate of emission increases to sixty a minute. Safely past the obstacle the number will again drop to thirty. The pulses emitted by some bats will have a central frequency of fifty kilocycles quite outside the receptive range of human ears. Bats, however, do give forth audible squeaks as well, and these, not unlike those emanating from mice behind the wainscoting, may also be heard as the bat flies overhead.

Quite a number of other creatures, ranging from *Steatornis*, the oil-bird, which lives in dark caves, to the whale, which is credited with an ability to use ultra-sonic sound for the detection of obstacles in deep water, are included in the list, and it is equally likely that some birds, as well as many insects, employ it also.

THE HORSE

The horse is a very highly-strung animal and its responses to stimuli are extremely rapid.

Its hearing is very acute and its earflaps move very freely in all directions. They point directly ahead when the horse is looking that way and using two eyes at one time. The two eyes of the horse converge (look forward in binocular co-operation) only when the horse is particularly interested in something in front of its body such as another horse, or a jump, or when it is keenly alert and excited. For the greater part of its passive life the eyes are directed *laterally*, each eye absorbing a different view simultaneously, occasionally taking a look *behind* the animal's body. It would be quite easy for a horse, walking along a country road, to watch a flock of sheep with the right eye and a man driving a tractor in a field on the opposite side of the road, with the left eye. It could then switch *both* eyes forward (binocular vision), take a quick look at an approaching motor car; then having drawn carefully into the correct side of the road, return its gaze to the sheep and the tractor.

The horse, in common with many of the ungulates, has a

poorly developed ciliary muscle which is not very capable of focussing the eyes. Instead, the horse depends more upon fixed focal lengths, which it selects by raising, or lowering its head.

The eyeball of the horse is not regularly spherical, so that the retina, the sensitive receiving membrane which occupies the concavity at the back of the eye, is further from the lens at some parts than at others. It is by raising the head high that the horse permits rays of light, reflected from an object close at hand, to fall upon the appropriate part of the retina, to give a clear picture. At the same time, the breadth of the horse's forehead between the two eyes is such that the eyes cannot converge onto an object less than four feet in front of them.

It follows therefore that when a horse approaches a jump, it must (if it is to see it clearly) keep its head raised. When it comes to the taking off point, four feet from the jump, it can no longer see the obstacle with both eyes so it tips the head a little to one side and observes it with one eye at a time.

In order to do this effectively the horse needs a free head, and as the amateur is probably hanging onto the reins for dear life at this stage of the proceedings, the prospect of either horse or rider reaching the other side of the jump safely is none too hopeful.

Horses have fairly good distance vision, (with head held level with the body) and can see other horses in a field up to two hundred yards away, and can hear them and be aware of their presence at even greater distances.

Horses do not show any remarkable sense of smell, although some exhibit a dislike for that of blood or warm, or decomposing flesh, and they are particularly sensitive to smoke. But it is likely that the dislike a horse shows to entering a slaughterhouse is not in any way psychic, as is so generally credited, but because the place is usually dark and noisy and gives forth an odour of flesh, and in addition it may retain a smell of gunpowder. Horses are naturally afraid of explosions including gunfire. All these add together to make the horse uneasy, but in spite of everything, one often finds a horse and cart doing regular work in and out of the abattoir and the horse has become completely conditioned to all the things I have mentioned.

One thing is quite certain. Any fear the horse may show of a slaughterhouse is not based on knowledge of death, or fear of a dead carcase, whether in its skin, completely divested of its hide, split in halves, or otherwise mutilated, lying upon the ground, or hung up by its heels.

No animal below man has any knowledge of the meaning of death. Animals which come unexpectedly upon their own friends lying dead, show nothing more than a passing interest. A dog smells its own brother, or mother lying dead in the kennel. The dog's impression of either is an up-and-standing one, moving about, sniffing and producing sound. The motionless figure does not fit in with any pattern of the animal as the dog knew it in the living state. Therefore, it cannot convey the information that the object before it was once a dog, but is so no longer. No member of the animal kingdom below the Primates (and few of these) knows that a body can cease to live.

The corpse has no more meaning for the animal than a sack of coal would have. A female monkey will carry and nurse her dead baby until it becomes mummified.

Horses have excellent taste discrimination and it is difficult to persuade them to take substances in their food for medicinal reasons even when these appear tasteless to us. Yet they show little dislike to certain drugs of a saline nature, such as Epsom salts.

THE NIGHT PROWLERS

Whether animals shall move about by day and sleep by night, or if they shall prowl by night and sleep by day, has been decided for them during the stages of their evolution. The decision demanded that each animal should live under the conditions best suited to its conformation and physical characteristics, and under those most likely to enable it to find food for itself and its family.

The owl is a bird, slow flying and low flying, at the mercy, by day, of all the lesser birds which harass and chivy it (but the terror of them all by night).

One may wonder whether the owl developed its pecularities of sight and wing and digestion, simply in order that it might be ideally suited to night prowling - or, if the pecularities of conformation and performance came unsought, and the owl decided to fly by night because it found itself physically incapable of living a diurnal life?

Once having become conditioned to the environment, does the law of survival of the fittest come into operation so that successive generations shall become more and more capable of survival within this environment? In other words, do animals develop characteristics to suit the circumstances, or do their characteristics force them to seek an environment in which they can live, if their species is to survive?

The owl subsists largely on other night prowlers, on those lesser animals which choose to walk by night because they are incapable of defending themselves in daylight. They slink about under cover of darkness, because at that time most of their enemies (but not the owl) are fast sleep.

Every individual that has survived until today, whatever its species, is a specialist in its own line. It would be reasonable to believe that it requires as much concentration, and nearly as much intelligence, to maintain an existence as a solitary, wild animal, as it does to remain alive within a civilised human circle.

Animals living in herds and other communities, as we shall read later, have less difficulty in earning a living, since when working on communal lines all flourish, or all starve. When famine appears imminent it is easier for a herd to force its way by weight of numbers onto a fresh feeding ground than for any single animal to do so, unaided.

When food runs short the mouse and the vole are left stranded, and by a kindly dispensation of Providence the hungrier they become, the more they are forced to wander afield and expose their bodies to the attack of predatory birds, which increase in number in direct proportion to the number of appearances made by the mice and the voles.

Birds which live upon seeds and fruit are compelled to adopt a diurnal existence, and in nearly every case birds other than the owls and nightjars are blind from shortly after sunset until sunrise.

The foxes hunt by day and by night if need arises, but they do most of their prowling by night and are particularly active during the hours of morning and evening twilight. Foxes are quite aware that poultry are easy prey by night unless securely locked away, and in this instance 'securely' is the operative word for the brain of the fox is quite capable of lifting latches and opening windows, unless they are very carefully planned. A tame fox belonging to the author discovered the way to open a car door, merely because he preferred sleeping inside the car rather than in a kennel inside the garage. He could also turn a door handle, so it became necessary to secure the pantry door with a padlock and key.

We can divide animals into groups according to their working hours:

(a) The *diurnal* animals, which operate during the hours of daylight and sleep by night.

(b) The *crepuscular* animals, which are most active during morning and evening twilight - shortly after sunset and around dawn.

(c) The *arythmic* animals, which are able to hunt during any part of the day or night; almost equally able to find their way in daylight or in nearly total darkness.

The foxes and the badgers and the night birds are mainly arythmic in their habits but as I wrote above, they are inclined to adopt crepuscular methods unless driven by hunger, or weather conditions, to operate throughout the night.

Most of the ungulates, horses, cattle, sheep, goats and deer, for example, are arythmic and feed through the night at intervals, alternating with periods of sleep. This is a security device as the whole flock or herd never sleep at the same time, and so sentries are always on duty.

Among the animals which are active throughout the night are the bats, the lynx and opossum, and the nocturnal foxes, which include the Fennec. In addition there is a vast number of small animals such as the Tarsier, found in Borneo, fitted with huge night eyes. Many of the night monkeys come into this category. The crocodiles are also most active by night.

The cat family, ranging from the lion down to the barnyard tabby, are arythmic by nature and although they move about by day and catch and kill during the hours of daylight, they are equally capable of hunting by night and sleeping the greater part of the day. Their eyes are sensitive to ultra-violet light and make good use of pupillary contraction and of the third eyelid during the hours of daylight. The domestic cats will sit by day apparently gazing at the sun, but if one takes a look at their pupils it will be seen that each is reduced to two pinpoints at the extremities of a slit. Yet in even a moderately dim light their pupils will be fully dilated over almost the whole of the visible part of the eye. Cats can use their eyes when the pupils are contracted as though they were pinhole cameras, which need no focussing.

Most of the modern bony fishes - the Teleosts - appear to sleep and to move about at alternate intervals both by day and night. Some observers query the possibility of fishes sleeping at all since they have no eyelids and cannot close their eyes. But a fish is capable of lying motionless for hours at a time with only an occasional flip of the tail, or fin, which is a purely automatic action not associated with consciousness.

It is probable that the mind of a fish can ignore visual impressions (provided it has any) and rest in a state of sleep with the lateral line alert, able to give warning of a stranger's approach.

Nocturnal vision is always conducted at short range, for the eyes even of a cat or an owl cannot see far in darkness unless it is illuminated somewhat by the moon or stars. While most of the night prowlers have such very large eyes, the one creature who shows up least by day has the smallest eyes of any animal in proportion to the size of its head. This is the mole, which has a perfectly-formed eye, but so tiny that it is practically invisible from the outside and comes away in the skin when this is removed. This fact has given rise to the story that the mole has no eyes. Such as they are, they are only capable of enabling the mole to find his own heap of soil.

COLOUR VISION

In this volume we are concerned only with the implications of colour vision, and not with its mechanism, which I have discussed in *Animal Vision* (Herbert Jenkins Ltd.) We most certainly are interested in the influence which colour perception may effect upon the mentality and behaviour of the animal, and whether it is able to put its perception to any practical use. It seems very unlikely that animals have any aesthetic regard for

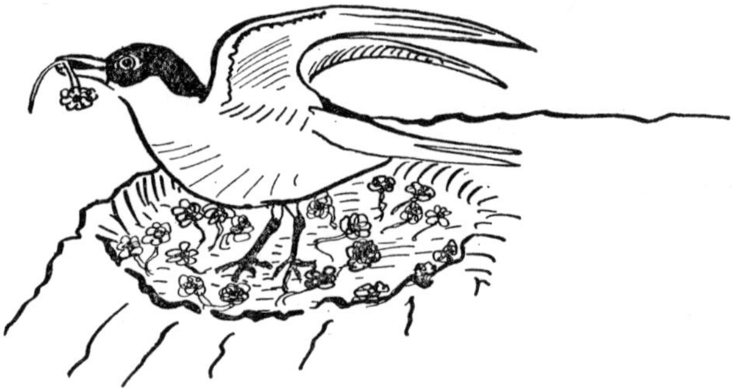

colour, although certain birds use flowers with which to line their nests. It is probable that the fact that the flowers have colour and are pleasing to the human eye is mere co-incidence.

I mentioned earlier in this book that many birds appear unable to distinguish between cocks and hens at a glance, even when the cockbirds are brilliantly coloured and the hens plain. *We* know the cocks are the birds with the coloured wings because we have been *told* that they are. If we saw the birds for the first time, without having received this information, we would have needed to watch their behaviour for a while before we realised that the differently-coloured birds belonged to the same species, and that all the coloured birds were cocks, and all the plain birds were hens.

In spite of all our vaunted intelligence we might not prove ourselves capable of getting very quickly off the mark. It would be less easy for the bird, since it would be unable to compare its own colouring with that of the other birds, and in any case its vocabulary does not include anything which differentiates the sexes objectively.

Let us consider what advantages may be derived from colour-consciousness by animals devoid of an aesthetic sense, a love of beauty for itself alone.

All of us have viewed cinema films, both plain and coloured. Perhaps it is because the colouration available in cinema films is still somewhat forceful or gaudy that all of us are not enthus-istic regarding their superiority over plain films in black and white, with an overall grey effect.

Even if we were left ignorant of the colours of the frocks worn by the ladies we were not unduly concerned, since we saw all the tone variation we needed, and quite forgot, as we watched, that in ordinary life we should have been viewing colour.

In actual fact there is no such thing as colour, any more than there is a physical entity brightening the world, which we refer to as "light."

Colour is only an impression, an adjectival one, qualifying the image reflected onto one's retina by means of light rays and translated by the appropriate brain centre into a form of consciousness which we term "vision." In order to get the benefit of colour the animal has to have a special kind of retina, packed with cones, and a special power of reception, capable of perceiving something additional to line and shape and greyness.

We are aware that the human brain can become conscious of about one hundred and sixty shades of colour in bright sunlight, and that the whole collection can be produced by mixing together red, green and violet, in their correct proportions. We are also aware that certain animals possess all the machinery one associates with colour perception but lack the ability to translate the image into anything more than greyness. The domestic cat has splendid eyes, the kind which incorporate all the mechanism required to give perception of colour, complete with the right kind of brain to appreciate it, and we are also aware that of all animals none could be more truly colour-blind than the domestic cat.

The dog is colour-blind and so, too, is the horse, as well as all the other four-legged animals. Only man, the higher apes, some monkeys, some fishes, and a lot of insects, can perceive colour. Although a good many animals recognise colour as distinct from brightness, no one can be sure that any specific colour looks exactly the same to, let us say, a bird and a man - nor can we be completely certain that a colour two men call green, or blue, produces an identical mental impression in each subject.

One of the remarkable things about colour in animals is that frogs and fishes (or quite a number of them) camouflage themselves by adopting the colours of their environment. The stimulus passes through the animal's eye. Yet, a flatfish, lying on the bed of the sea with eyes directed upwards can have very little, if any, visual impression of the colour of the mud or sand upon which it is resting. Nor can the flatfish see its own body and decide whether at any moment it is black, green or grey.

A mackerel can catch a glimpse of very little but its own tail, and even if it can see a lot of other mackerel there is nothing to suggest that it regards any one of these as being the facsimile of itself. The brains of mackerel do not permit any such comparison.

Then, the beasts of the field, the antelopes, the deer, and the giraffes, to mention only a few, wear colours calculated to make them invisible to the animals which would prey upon them. Their colours merge into those of their surroundings and it is presumed that the animals which would like to eat them will now be unable to see them.

It is not to be presumed that this form of camouflage has originated through any consciousness on the part of the animal, since all those I have mentioned are completely colour-blind, including the animals which would prey upon them.

One can only conclude that it is not colour which makes the camouflage effective but *tone*. If we photograph these animals against their normal background we shall find that they are far less conspicuous to our own eyes when we see them in a print on tones of grey against similar tones of grey than when we see them under their natural conditions, made even more conspicuous by vivid colouring. It is clear that the actual colour was never intended as a disguise; instead it was the matching of *tones* which made the camouflage effective.

The effect was intended to deceive eyes which *did not* perceive colour. It is only by chance that *we* see them in colour, and we and the apes are the only animals that can do so. The apes will not hurt them, but we humans, with our highly developed intelligence, may well do so.

During the course of evolution those animals which grew coloured hair on certain areas of their skin surface managed to survive and passed their genes for hair growth and colour onto their descendants. Those which grew hair of the wrong colour then developed some variation in the blackness, or whiteness, or greyness of their bodies as seen by the predatory animals. If they became conspicuous they were too easily visible and met their fate. In this way only those which retained their invisibility lived to propagate their kind.

Although birds see colour, they do not discriminate nearly so well as we do. Pigeons do not see violet and farmyard poultry do not see blue very well, if at all.

Those readers who are interested in the colour sense of bird should refer to Dr. Stuart Smith's, *How to Study Birds*, in which a number of experiments are described in detail.

Horses have no colour consciousness but they are aware of differences in brightness and so can distinguish between a grey horse and a black, or between a brown and a chestnut.

Dogs also distinguish shades of grey but cannot distinguish grey from blue of a similar degree of brightness.

All the apes, including the gorilla, the chimpanzee and the ourang outang, are good at colour discrimination, and many of the monkeys; the Rhesus, Diana, Green monkey and Capuchins in particular; but it is rather doubtful whether the marmosets and some of the lower primates have a great deal of colour sense.

During the breeding period the large seat patches of many of the apes swell and become almost scarlet. The apes appear to recognise sex visually as we do, while all the other mammals rely upon smell, and behaviour patterns.

The birds fail to realise the significance of sex colouration in spite of their good colour vision.

One of my Rhesus monkeys, "Peter," of whom we shall hear more in a later chapter, would make violent love to a lady dressed in red and ask to be nursed and petted, but later in the day when she appeared in blue, he would receive her with screams of rage which made it necessary to lock him in his cage until she had departed.

The question of whether fish see colours has aroused a great deal of controversy, especially among those who spend happy hours selecting the colours of their flies and lures, and deciding what colour they will dye their maggots.

Experiments have convinced a great many workers that fish have a colour sense, while other experiments appear to have demonstrated equally clearly that they have none.

Sea fish do not see colours at any great depth, since in order to see colour one needs plenty of illumination.

The colours of flowers and houses, of clothes and curtains, all disappear in the dark, even when the objects themselves are clearly visible. There comes a degree of darkness when one can no longer decide whether a card is white or black (although neither white nor black ranks as a colour). In muddy pools and lakes no fish could see a colour at a greater depth than four inches. In open bays where the water is comparatively calm and absolutely clear and the sunlight streams through to a sandy bottom, fish can probably see colours far down in the water.

But artificial flies floating dry on the surface could reveal their colours to a fish gazing upwards only if the water were clear and the sun shone through the wings of the fly. Otherwise the fly, however brilliantly coloured, could be no more than a silhouette. Nor can fishes see colour at night, which is why anglers use a white fly.

According to Chapman Pincher (*A Study of Fishes*), their food preferences may be influenced by colour. Some fish favour blue or green and prefer dim blue to light red. Other fish select red or blue baits, either but not necessarily both. In an aquarium, fish tend to congregate in the brightest light, and if a spectrum is projected, choose the yellow or green bands. If the colours are progressively shut off from the violet they move through the colours towards the red end of the tank. When nothing but red remains the fish are blind and refuse food, but commence to feed once more directly the light is changed to blue.

The cartilaginous fish, the sharks, rays, and the like, are genuinely colour blind. It has never been definitely proved that *any* of the bony fishes do *not* possess colour vision.

The eyes of insects, devoid of cones and rods, are often capable of discerning colour, though what they see may not be the same as we see. Few are able to see red, and it is likely that it is the smell of ripe apples that attracts wasps rather than their colour.

One thing appears certain. With the kinds of minds that animals possess, very many of them would have been more greatly handicapped than they are at present, if they possessed colour consciousness in any way approaching our own.

ABOUT OURSELVES

We humans possess remarkably simple eyes but they are coupled with a highly efficient brain. The result is that we are favoured with a quality of vision, only exceeded for distance, perhaps, by that of some of the predatory birds.

We have a fairly good tactile sense largely centered in our finger tips and tongue but with all due respect to the difference in brains, it is doubtful whether ours will compare with that of the bees and ants.

Our hearing is extremely variable even within the limits of normality, and this is particularly the case in people under thirty years of age. The hearing of most animals is better than ours, and dogs, cats and birds, together with most of the wild animals hear much more clearly than we do.

In domestication dogs and cats suffer a great deal from parasitic invasion of the ear passages, which frequently induces early deafness if neglected.

Both the insects and the birds hear a great deal of sound of which we are not conscious, and both dogs and cats hear supersonic noises proceeding from certain animals.

Our olfactory sense is very poor, and when we realise that almost any dog in an obedience class at a show will recognise a handkerchief which a judge has handled, out of a half-a-dozen others, sometimes while the dog trots along the row, we begin to discover that the world of the dog must be choc-a-bloc with smells of every degree of concentration. It is almost impossible for a man, with so poor a sense of smell, to imagine the part that this vast variety of perfumes must play in the dog's life, far exceeding that associated with any other form of perceptivity.

Although the lower animals may not understand a great deal regarding the origin and meaning of many of the smells they encounter, quite a number of them *do* have meaning, and these, acting as stimuli trigger off responses which may vary from simple salivation to a desire for battle, or an urge towards mating.

Impressions and perceptions, such as these, form the landmarks in the animal's pattern of living and they play the same part in maintaining interest in its life as some of the man-made entertainments do in our own. We have our T.V., our papers and our books, we listen to the news and to all the local gossip, because our ears can tell us more than most of our other forms of perception. We can exchange ideas and occupy our minds with thought and the exercise of a vivid imagination. The dog has to make do with his nose.

Visual Perception and the Gestalt Theory

It would be quite misleading to imagine that the world as we see it, and the world as it appears to another kind of animal, are pictorially identical. Nor must we believe that objects which the human eye would single out from the confusion of half-seen images as being potentially interesting, would be the same images that would attract the attention of a dog, or an alligator. When we discover what the objects are which appeal to any particular animal we shall begin to understand the nature of that animal's little, private world.

In the Universe there is an almost incalculable diversity of living creatures, of all shapes and sizes, of varying patterns and degrees of solidity. Each and every one of them has its own scheme of living and its own special relationship to the world around it.

The number of objects which will attract the attention of a kangaroo or a grass snake, will be far less than the number of those which will interest a Rhesus monkey or a chimpanzee, and the number attracting the chimpanzee will be considerably smaller than that which will excite the interest of an educated human being.

Each kind of animal is concerned with the necessities of life; food, shelter, and reproduction.

Man, being at heart an animal, has an eye for all these things, too, but his mind is better furnished, and his outlook is much broader. He has to find exercise for his active brain and his imagination, and he demands something more from life than the mere satisfaction of his needs. He is ever on the lookout for fresh thrills, excitement, and all the trimmings that help decorate his civilised existence. He will notice a great deal more in the panorama which moves past his eyes than any animal ever would, and he will make a great deal more investigation into the items which stir his curiosity. He has imagination and ability to plan ahead, and so he seeks not only things of immediate importance but those, too, which may play some part in his life at a later date. No animal, other than man, can do this.

The great majority of animals desire only to eat, sleep, reproduce, sleep, eat, in that order, *ad infinitum.* Animals, watching the world go past, notice only those things which have for them functional significance - something to eat or to chase, somewhere to sleep, other animals, the roads and paths they travel, and the visual cues associated with simple pleasures.

There is a close similarity between the mind of the human infant and that of the animal but while the mind of a baby develops, that of the animal after attaining a certain limit, stands still. During the first few weeks of a baby's life it sees practically nothing and can ascribe no meaning to anything it does see. It notices movement, and as the objects which move in an infant's presence are usually human beings it begins after a few weeks of life to distinguish animate from inanimate objects. Not too successfully perhaps at first, for the baby of four months may commence to greet its mother with a smile. The joy of the mother may be less evident, as I pointed out in Chapter Two, when some tactless scientist assures her that her baby will smile in precisely the same way at a turnip mounted on a broomstick. At the age of six months the objects within the infant's range of vision form a confused mosaic pattern, of which the elementary parts are still neither separate nor distinguishable.

Moving across the mosaic there may sometimes be other and smaller fragments, which have become dissociated from some part of the general pattern and are now establishing a constantly changing association with the remainder of it.

These moving portions hold the attention to the exclusion of all else, just as we, ourselves, gazing at a starlit sky, lose all view of the general pattern of the stars in the heavens as we watch a blazing meteorite making its way across the expanse of sky.

The response of a child or of an animal must be very similar. It takes the form of a general visual pattern, the elements of which are not separated, but are considered only as parts of the whole, and in accordance to their relation to the whole pattern.

In the child, as in the puppy, maturation and learning gradually increase the ability to distinguish simple elements without any special regard to the rest of the picture, just as a bee flying a few inches from a wall may ignore the patterned wall paper as a whole, but may be attracted at regular intervals by a small pink flower painted on the paper, which to our eyes may form no very distinct part of the total pattern.

The brain of the child will mature and continue to develop and although the puppy of a year and the child of two years may be approximately on an intellectual par, from now on the child will advance at an even faster rate, while the mind of the puppy will stay almost at a standstill.

How much any one of us sees of the world around us will

depend upon our dreaminess, or our concentration, at the moment when things pass us by. A great deal of our future programme will depend upon our awareness, our knowledge of what certain portions of the mosaic imply and whether they give rise in our minds to further imagining.

A dog walking down a street pavement, or sidewalk, sees within the mosaic, shops, nameplates, buses, cars, people and perambulators - but they convey nothing to his intelligence. *We* know that the shop across the road is occupied by Mr. Jones, that he deals in washing machines and refrigerators. We know how much he charges for them because we can read the price tickets in the windows. We may know Mr. Jones and all the Jones children. We see the 'bus go by and we can see on its plate where it has come from and where it is going. We are acquainted with the lady pushing a perambulator past the shop and we look at her baby and remark on how fast it grows.

And so we observe separate items in the mosaic with interest, while a dog gets only the general view, unless by chance something "doggy" excites his attention.

But it is not always like this. When worried, absent-minded, or wrapped in our own thoughts, we may stride from end to end of a busy street, seeing nothing and nobody, and if we do take note of the scene we visualise the complete mosaic without picking out of it a single item for further elaboration.

The dog knows nothing about shops, why they exist or what they trade in. He cannot read the signboards nor can he see the objects displayed in the windows. Even if he could he would be entirely ignorant as to what they were, what they were used for, and why they were in the window. Nor would he care.

In the wilds a similar state prevails. The animal sees the whole picture and picks out only the odd item which has a meaning. It notes the presence of an enemy and seeks something that looks like food when it is hungry. Nothing else; the trees, the hills or the sky; are of any importance whatever, and they make no addition to, or demands upon, the animal's intelligence.

In the house, furniture means little to the dog or cat. A

washing machine is merely an obstacle, the kitchen table is some form of roofing, but either dog or cat will know the whereabouts of the pantry, or of the cupboard in which food is stored. The world must be cluttered up in the animal mind with completely useless objects, not worth a glance - but if anything moves within the visual mosaic that animal will spring to immediate attention.

My Scottie, sitting on the dining room mat in front of the fire, oblivious apparently to everything, never misses a sweet thrown to him from halfway across the room, entirely without warning.

And yet, when he walks into the same room of an evening preparing for a quiet snooze before the fire, he will stand gazing at an envelope nestling against an ornament on the mantelpiece, which was not there last night. He will look at the clock which is not in quite its usual position, and while gazing at these intently, he will bark to attract my attention, and not until I have made an inspection and assured him that everything is in order, will he settle by that fireside.

What can be the reason for this unusual demonstration - why this increased ability to observe something unusual?

I think the answer is that the street outside is common ground and of no particular consequence. The room in which he spends his evening is his own special territory, to be watched and protected. It is only right that a pack member should report any discrepancy to his pack boss!

The Kinaesthetic Sense

This relates to sensations which give notice of bodily position and movement. The receptors lie in nerve endings situated in the muscles, tendons and joints. Stimulation of these receptors conveys to us the knowledge that we have hands and feet, fingers and toes, limbs and a body, and even if we were blindfolded, or lying in a pitch-dark cave we should still be aware of our bodily construction by virtue of this kinaesthetic sense. It is by its aid that we recognise that objects have weight, and a dog recognises the resistance of the earth when he wishes to bury a bone.

Pain

In our everyday life among everyday people we meet two schools of thought regarding the sensitivity of animals to pain.

The one affirms that animals do not suffer pain as acutely as we do - the other is sure that animals are far more sensitive and feel pain more acutely than we, ourselves, can do.

So far as skin sensation goes, the majority of animals appear to be as sensitive to pain as the average human being. It is not so easy to know when an animal is in pain since animals are more stoical than men and less inclined to exaggerate or anticipate.

A few animals may be rather less sensitive in parts of the body. A bull may have a skin an inch thick over the neck, back, quarters and buttocks and thinner in other parts. Thick skins are usually less sensitive than thinner skins and the bull usually makes little fuss over a hypodermic injection, but as one never sticks needles into a bull's skin for therapeutic reasons until the animal is well under control it may be difficult for the bull to express his real feelings.

The skin of the horse is highly sensitive to any sort of external stimulus and therefore horses react very markedly to needle pricks. Horsemen control horses to a certain extent by using "the twitch" during such operations as shoeing, clipping and administering medicines. The twitch is a loop of cord applied around the upper lip and tightened by twisting it with a stick until it causes pain. Either the pain thus produced is greater than that of the operation - or the horse cannot move without causing itself increased pain, and it therefore remains still.

A third explanation lies in the belief, now becoming more prevalent, that pain in certain parts of the body is prevented by stimulation of another area. It is thought it may be possible to map out corresponding areas so that mild electrical stimulation of the skin of some portion of a limb, let us say, may produce an anaesthesia of the skin of the scalp or of some internal structure.

These are only random locations but will serve to point out the basis of the theory.

Another factor which operates in man, rather than in animals, is anticipation. A man will lie awake all night dreading a minor operation he is aware will be performed on the following day. The animal knows nothing about what is intended, and the operation may be over and done with almost before the animal is aware of what is being carried out.

Horses and dogs undoubtedly feel pain from a diseased tooth, or the formation of an abscess, as we do, but as animals go lower in the biological scale the degree of sensitivity is far less. A worm, cut in two with a knife, wriggles in reflex fashion, then both halves go their own ways. It is unlikely that the worm "feels" or experiences severe pain since its mental apparatus is not sufficiently developed to record the sensation.

Insects sometimes walk about when deprived of most of their internal organs, and the loss of a few limbs does not appear to produce any particular response. But some of the domestic animals appear almost equally insensitive. A pig will make little fuss of an extensive mutilation but will be unable to walk about after a prick in a foot. A bullock will continue to graze with a fractured limb. A small amount of blistering ointment rubbed into a horse's hock will drive it into a state of panic, accompanied by stamping and sweating, when the same application rubbed into a man's back might be followed by only a feeling of discomfort, which he would put up with quite cheerfully. Animals, however, withstand skin burning which would kill most human beings.

Men and women, as well as children, subjected to severe pain, may succumb to what is termed "shock." Horses and dogs both exhibit this condition when they undergo severe operations under anaesthesia, particularly inhalation anaesthesia, and the avoidance, or relief of shock is as important to successful surgery as the performance of the actual operation.

Cattle and pigs are least susceptible to shock, but sheep are particularly susceptible.

Chapter IV

ORIENTATION - HOMING AND MIGRATION

Every animal is especially related to the space surrounding it, as well as to the objects it may encounter during its passage through space.

Animals move in various ways. Men walk on the earth, birds fly in the air, fishes swim in water. The medium bearing the closest relationship to the animal differs in each instance.

Animals which walk make solid contact with the ground beneath them; those which fly alternate between the air and the earth, or offshoots of the earth, with which they, too, make intermittent contact. With one or two exceptions fishes spend the whole of their lives immersed in water.

All of these animals have their own special needs. Man, who stands upright, needs to balance his body on two legs and two feet. His relationship to the soil beneath him must be controlled largely by gravity which endeavours to bring him down to earth so that he measures his whole length upon the ground. But a centre in his cerebellum, combined with the semicircular canals within the internal ear, enables him to co-ordinate the muscles of his limbs, particularly his lower ones, so that he overcomes the tendency to fall and maintains himself in a standing position.

Birds have to maintain a normal relationship with the earth beneath them, even when suspended by their wing-spread in mid-air. They have to be certain they are flying right-way-up and not upside-down, which can happen very easily unless a flying animal can see other objects with which to compare its position in space.

They have to know, also, the line their flight maintains in relation to the earth beneath, so they will know if they are ascending, descending, or flying on an even keel. They need to exert sufficient control over their movements so that they will not fly into obstacles, including the earth itself, and they must know how to "apply the brakes" when they wish to land.

Fishes, as I mentioned in the previous chapter, possess accessory sense organs, one of which is the "lateral line" which enables them to recognise and avoid underwater obstacles. Also, they have internal ears which are not connected to the exterior, nor has the fish any external aperture, nor visible ears. Nevertheless these organs, such as they possess, enable fishes to maintain their balance in water, swim the right side up and determine whether they are approaching the surface or sinking into the depths.

A great deal has been written around orientation in animals in space, also concerning auditory orientation, as exhibited by birds, dogs, cats and other creatures which succeed in returning to their homes after absence elsewhere. In addition, even more attention has been directed towards a study of migration as observed in birds, insects and a variety of other creatures.

In spite of the extensive research carried out, nobody, up to the present time, can say authoritatively how dogs and cats find their way home over great distance, or how pigeons and a number of varieties of our wild birds return home after traversing unknown territory when they have been removed from their natural habitat.

Invariably one comes finally to the conclusion that some animals must be in possession of sense organs, or modifications of existing centres of perceptivity, of which we have no knowledge.

We are told by a number of observers that both homing and migration in birds are controlled in some unknown way by the internal ear and that the semicircular canals play a major part in this control.

We *do* know that a man or other animal with any diseased condition of the internal ear will suffer from dizziness and incoordination of the limbs, so that it becomes impossible for him to walk a straight line or preserve a sense of equilibrium. The cerebellum, the hinder part of the brain, also controls co-ordination of the limbs.

But when we announce that the proper functioning of the internal ear enables a bird suddenly to take off in flight and come to land hundreds of miles away across the ocean; that with its help an eel travels from our rivers to the Sargasso Sea without aid of chart or compass, we shall have to seek some other explanation.

The weight of experimental evidence indicates that the internal ear of the bird, and its semicircular canals, play small part, if any, in the orientation of birds in the air, whatever part they may play when they come to rest on land or water, or in the treetops. Homing pigeons in which the semicircular canals have been put out of action, show no loss of their ability to return to their lofts.

Whatever lies behind migration and the ability to find its home on the part of any animal, must provide an unfailing sense of direction, sufficient to enable a young bird, not a great many weeks out of the nest, to take off unaided, and fly in advance of its parents to a land predestined to fulfill all its needs - a land of which it has no knowledge whatever, and one which it cannot recognise when it finally arrives.

One might imagine that homing as seen in pigeons, and migration as it applies to many kinds of birds, are similar phenomena varying only in degree.

A pigeon flies home over a distance of twenty miles or fifty miles, let us say, while a swallow flies to its new home which may be two thousand miles away. It may take one hundred and nine days to make the journey at a daily rate averaging twenty three miles a day. Dr. Stuart Smith refers (in *How to Study Birds*) to the redstart which leaves its winter quarters in N. Africa after the main swallow and willow-warbler migrations have left that area, and yet by the time Denmark is reached, it has caught up with the main surge of these birds. The redstart's rate of progress is thirty-three miles a day in the West and fourty-one miles a day in the East. Migration over land is a steady drift reckoned in miles per day rather than miles per hour. But migration over seas and oceans must be a *continuous* effort; there can be no stopping half way although countless birds make for the lights of a lighthouse far out at sea, no doubt hoping to rest.

But homing and migration each have many distinctive characters. The homing pigeon knows its home and has become conditioned to it long before it starts professional flying. It receives training, flying at first only a few miles, the distances being gradually stepped up. The successful homing pigeon is the one which is able to profit by education. It is encouraged to develop a nostalgia; probably it leaves a mate behind and has an urge to return to its loft.

The migrating cuckoo, born in this country, with no knowledge of its own parents, may have to travel from Kent, or from Cornwall, to Africa, a land it has never seen and of which it has never heard. The pigeon may feel homesick but not the cuckoo, surely. Why does the baby cuckoo, become restless and crave for a land so far away. The cuckoo does not know that the available insect population in this country during winter is almost nil, there are enough to keep the adult birds for a few more weeks. Even adult swifts leave this country early, long before insects become scarce.

Perhaps one main reason why none of us can imagine why birds fly away at certain seasons, and how they contrive to display such an unerring sense of direction is because we ourselves are almost totally devoid of one, and therefore unable to understand how any other animal feels when putting its sense of direction into operation. Blindfold any man and turn him round three times on his own feet, and he will be quite unable to say before the handkerchief is removed which direction he faces.

Very few persons sitting in a room in a house they do not know, can say without staring out of the window, where the north lies. But if when thirty woodpigeons are busily feeding in a field, heads well down and facing in all directions, someone fires a gun, almost the whole flock will rise as one bird and make off immediately on a definite course which will carry them either to a roosting place or to another feeding ground. There is no panic and no disorder, as might happen in a crowded cinema if someone shouted "Fire!" It is seldom that more than two per cent of the flock will circle around but when they do so these birds will almost invariably set off together in an opposite direction to that taken by the main part of the flock. When we remember that the eyes of pigeons are only four inches above ground level and their view is often obscured by banks and hedges, this ability to get right away in so orderly a manner is the more surprising.

It is unlikely that animals; birds and insects in particular; experience at any time an unlimited range of freedom within the length and breadth of the area of locality available to them. From personal observation it seems more probable that small birds and many varieties of insects such as bees and winged flies, have their own central points of location; their homes, their nests or their favourite haunts; and that they move from these not completely at random but in more or less defined directions.

If one contrives to mark a number of woodpigeons in some way that will enable them to be identified in flight one will find that when flying solo each bird will follow a fairly regular line of country so that it passes over the same landmarks a number of times in the same day. Most birds have their regular lines of flight and follow these in much the same way as a rabbit uses the same track across a field or plain when travelling any distance from one place to another. When birds move in flocks it is usually from one feeding ground to another, and then to a roosting place, and the flight lines of flocks, although they differ from the solo flight lines, will nevertheless follow a fairly regular pattern unless the birds are unduly disturbed. Individual needs and flocks needs are not necessarily identical.

The tendency to make use of a particular tree, and of one or two flight lines, establishes the territory of a bird as faithfully as a kennel and a few yards of garden may constitute that of a dog.

The peregrine falcon discovered the habit which birds have of developing recognised lines of flight long before it became evident to mankind. The wildfowlers are well aware of the habit and make good use of their knowledge on the saltings and marshes. They know full well that when a bird has developed one type of directional behaviour it will continue to make use of it - unless the habits of mankind arouse suspicions in its mind. When a feeding place becomes deprived of nearly all it has to offer the flocks will move elsewhere, but individual birds will continue to visit it.

Some psychologists speak of a "group memory," which they consider has been handed down from the bird's ancestors. While newly-learned patterns of behaviour may influence the behaviour of an animal's descendants in the sense that only those which acquire such behaviour survive, it seems unlikely that a group memory should be transmitted in this way. A tendency to fly north-west or south-east may be and possibly is transmitted to the off-spring but memory hardly comes into the picture, especially as there are no facts to remember. The bird is unconscious that it is doing anything unusual, it does not realise that it is migrating or even consider the necessity of doing so, nor does it know for what locality it is heading. Nor does the bird know from memory which direction is south-east. If it is impelled to fly in a given direction it is through the activity of some inner force, which certainly is not memory.

If group memory were a transmissible feature, a great many children at five years of age would recite the Lord's Prayer spontaneously, announce that William the First conquered Britain in 1066, and that a person named Columbus was *not* the discoverer of America.

In what is sometimes termed "local orientation," both sound and scent play a very great part. Flocks of birds, for example, go through a regular drill, they fly and turn in unison by watching the wing-tips of their neighbours in flight. Other varieties, notably sea birds and waders, rely largely on call notes, some of which are so high in pitch that our ears do not register them.

Auditory orientation is almost as important to the bird as visual aid. It recognises the direction from which the sound has come, its volume, the expressional tone in which it is uttered, and especially its emotional character, whether it expresses alarm, suspicion, or if it be the serene and melodious twittering of waders congregated safely and happily on a sand bank.

THE HOMING OF BIRDS

It must not be forgotten that although the most frequent indications of homing ability is seen in birds, it exists also in dogs, cats, bats, mice, squirrels, crabs, and many other animals.

A great many theories have been formulated (and in the main part subsequently discarded) in order to discover some feasible explanation as to how homing pigeons, sent by train, packed in crates, a distance of say a hundred miles, find their way back to their lofts after being liberated. Perhaps one should say that although a large percentage return, not all do so by any means. Some birds are shot, or killed by peregrine falcons on their way, while nowadays they may come down to feed and be poisoned by crop sprays. Some get completely lost and fly off in an entirely wrong direction and a few make no attempt to fly home but settle down in their new locality or close to it, either as strays, or as additions to local lofts.

On the whole, the greater proportion of birds do arrive home sooner or later. There is a great deal of variation in pigeon performance, even at its best.

Viguier and Thauzies declared that the pigeon orientates itself by the magnetic field within the earth over which it flies. Experiments have shown that pigeons cannot be deflected or guided either by magnetic fields or electric currents.

The retracement theory sponsored by Reynaud likened the mind of a bird to a gramophone record which could be played again in reverse. This would imply that the bird travelled abroad and back again later by an identical route. Quite often birds are sent away from home in boxes or "travel blind" in railway trucks from which they receive no impression of the route or direction they travel. Birds may return quite successfully by quite a different line of flight.

Another common belief is that the pigeon circles on high, picks out an object it knows, flies towards it then rises again and looks for another landmark. Since many pigeons are on a "first time" flight over the particular area and could have no experience of the landmarks, this seems an unlikely explanation.

Resident starlings captured from different localities, ringed or marked with paint, have been taken in boxes, carried below-deck far out into the North Sea and liberated there. These birds return to their homes with greater celerity and certainty than that shown by pigeons liberated on land.

Curiously, modern pigeons are not provided with the best vision since they have usually lost the two foveae, the points of acute vision in the retina, although these are present in a great many varieties of wild birds, especially the hawk family and the insectivorous birds.

Pigeon fanciers who breed and fly homing birds pay a great deal of attention to what they call "the eye sign." This term refers to a rather indefinite appearance which seems to indicate a slight haziness or iridescence of the iris around the periphery of the pupil, seen when the latter is dilated. Breeders assert that birds showing the "eye sign" are the best homing birds and accordingly they select these for breeding stock. So far as can be ascertained the eye condition can exert little salutary influence, unless by chance the appearance is present in particular strains which happen also to be good at homing.

There is a difference between the performance of birds which simply return to their homes and those which migrate to foreign lands, or to different parts of the same country. Many of the latter are compelled to fly frequently by night as well as by day, while homing pigeons are liberated so that they may travel by daylight if possible. Quite a number are trained to fly only in one direction until they have gained experience; they acquire knowledge of the country in this way.

The deciding point must be whether the bird starts off heading in the right direction, or is capable of changing its course before it has gone a great distance. No doubt the practice of liberating a great many pigeons at one time helps the young and inexperienced birds as the "old hands" will probably set off in the right direction which will enable the less gifted to follow them.

THE HOMING DOG

Quite a number of dogs stray away, possibly on a courting expedition, and seem unable to find their way home again after all the excitement is over. Against this there are innumerable records of dogs which have been transported great distances from home, sometimes several hundred miles into completely unknown country and yet have returned sooner or later to their own homes. In some instances the dog has moved with its owners into a fresh part of the country and the dog has shown greater love for its old home than desire to stay with its owners. In others, the dog has been stolen and taken away usually by car, and has subsequently escaped and made for home. It is exceedingly likely that the majority of dogs which find themselves deserted in strange country, fifty or a hundred miles away from home, never get back at all.

Advertisements for lost dogs in newspapers are numerous enough to make this seem probable, but their number may be no guide as most advertisements are inserted promptly after the dog's disappearance, long before it would have time to make the return journey.

Then again, stray dogs are never welcomed in sheep and cattle raising districts. They must eat if they are to remain fit to travel and sometime they must sleep. They may steal or beg, and in either case it is seldom long before they become noticed and probably captured. In country police stations in Britain dogs are seldom kept for more than a week before they are destroyed, while in many countries where rabies is prevalent a stray dog may be shot at sight. Dogs, too, that are forced to travel constantly soon become foot-sore and weary, and many dogs that return home after a week or a fortnight's marching lose the greater part of the pads of their feet. All this makes against the chances of success especially when a lame and foot-sore dog is compelled to travel on roads carrying a great amount of fast motor vehicles. I think it may be taken for granted that a far greater number of lost and stolen dogs would return to their homes, if they could be left alone, unhindered in their attempt.

Few dogs have any chance of escaping the law when once apprehended, unless provided with a collar bearing the name and address of the owner. Thieves make sure that they do not wear this, and even those dogs which are captured and tied up and escape usually do so leaving their collars behind. It thus becomes apparent that the lost dog not only has to compete with the difficulties of road navigation but it also has to contend with the ill-will of the human population it may encounter on its travels.

Just as there is a difference between the homing pigeon and the migrating bird so one exists between the homing dog and the migrating bird. The bird departing on its first migration (which is as successful as later trips) has no knowledge of where its destination may lie. The dog is hard put to it to find his way - the bird cannot go wrong.

At the very moment of setting off, the dog needs to know his main direction. It is useless to start off at due north if its home lies south-west of the starting point.

Most birds fly south-east or south-west in September and retrace the same route in April or May. The directional guide, be it what it may, is inherent in the bird, fitted somewhere into its mentality and highly dependable. The dog seeks its home purely from nostalgia. The bird migrates because at the particular season of the year its need to fly has been stimulated by the pouring out of a hormone into the bird's blood stream.

Man has lost his directional instincts as the result of the kind of life he now lives. Perhaps a million years age they were more pronounced, we shall never really know.

He is still conscious that he may proceed in one of four directions. He tends to walk forwards, can reverse at need, and is very conscious of the difference between right and left, unless he happens to be ambidextrous. Right and left handedness may be the sole relics of his directional sense. These characteristics depend upon an uneven development of centres in the two sides of the brain.

Animals are usually ambidextrous and it would be interesting to know whether the sense of direction is more marked in these comparatively rare specimens of humanity who also use both hands equally well.

Some experiments in animals have been carried out to determine whether four-legged kinds move off with the right or left fore foot. These experiments appear to have little value since it did not appear to be properly understood that few four-legged animals have no choice in the matter.

When a horse moves off from a standstill its two forelimbs are "locked" at the elbow joint, so that if it is standing up squarely on all four feet it is unable to advance either forelimb until it has first flexed a hock and allowed the weight of the body to be taken on that particular hind limb. As soon as it has done this it can flex elbow and knee of one forelimb, usually the opposite one to the hock it has first flexed. Which forelimb advances first must depend upon how the horse happens to be standing at the time of the order to move forward, whether the front and hind halves of the body are level, or one higher than the other. It will depend also upon the contour of the ground on which the horse is standing.

Dogs and horses which shake hands are guided by the trainer, the position in which he stands, and whether he faces slightly to the right or the left of the animals.

Horses are easily taught to lead off with either leg at the canter, and to change the leading leg during this gait at the will of the rider.

Apes and monkeys are almost invariably ambidextrous.

A great many experiments have been made from time to time in which dogs were deliberately "lost," and their progress watched in such a way that the dog did not know it was under observation.

It would appear that the majority of dogs after a little preliminary skirmishing, set off on a line roughly approximating the direction in which their homes would lie. If the dog strays off its correct course it seems to become uneasy, its rate of respiration increases and the dog looks worried and potters airmlessly around for a short while. Sooner or late it appears to sense the correct line once again and its demeanour changes to apparent tranquility so long as it is following the right line.

The dog appears "comfortable inside," as one observer put it, when travelling in the direction of its home, but mentally uneasy or again "lost" when it takes a wrong turning, or when the road takes a great many turns and twists.

When observing the behaviour of the dog in its homeward travels, one cannot help thinking of what happens when one tunes a portable radio set into a station and rotates the set upon a table. If one carries the set by the handle, turns around until one gets the loudest signal from the set, then paints an arrow along the top of the case pointing straight ahead, one will be following the line indicated by the arrow whenever the volume is loudest, even if one were blindfolded. If the arrow happened to point towards one's proposed destination one would be travelling all the time in the right direction so long as the maximum signal volume was maintained.

Although this is only intended as a simple analogy it would almost appear that a dog or a bird must sense something comparable with this, even if one substitutes mental distress in the dog for a dimunition of signal strength in the case of the radio set. The main difference here is that the transmission comes from the dog in the one instance and from the "home" station in the radio apparatus. Whatever it may be that indicates to the dog that it is travelling in the right or wrong direction, comes from within the dog rather than from some outside agency.

If man ever possessed a sense of direction of any great value, socialisation and civilisation have made its use obsolete. It is no longer necessary to mankind but it is still essential in the case of the dog, the pigeon and the migrating bird. A man need only look up a map, and reach for a time-table. His sixth sense, if it still exists, would not help him to work out his programme of travel today.

One is reminded also of the mosquito when considering direction finding in space. The male and the female mosquitoes each carries antennae of the bottle-brush variety which act as aerials, those of the female transmitting radio signals and those of the male receiving them over considerable distances. When the female's bottle-brush commences to vibrate the male turns himself about until he finds the direction from which he receives the best signal; then he tunes in on this wave and follows it until he meets his prospective mate.

It may be that the dog can produce something of the nature of a radio wave which can be directed towards its home location. It is equally likely that the true explanation is something entirely different.

When we aim at becoming psychologists we must always presume that there is some far more simple explanation - if only we knew what it might be!

The result of experiments carried out in animals which have been conspicously marked either by paint or by banding, has demonstrated that quite a number of wild animals other than birds, have the power of determining direction. These include mice and rats, hedgehogs and bats.

Horses and most other ungulates do not appear to have a great homing ability in strange surroundings, though horses which have been hunting over a lengthy run will sometimes return home alone after parting company with their riders.

Using wild mice which he had caught in traps, Stickel of the Patuxent Research Refuge in Maryland, released a number of these at distances of five to seven hundred feet from their homes. A mouse seldom travels more than fifty yards from its hole. Out of these, eighty two per cent of twenty seven males returned almost at once to their original homes. One went for a considerable distance in an entirely wrong direction and two failed to return at all. Sixty-two per cent. of twenty-one females returned and six failed to return.

Pipistrelle bats possess a homing ability over several miles and some of the European pipistrelles scheduled twenty-five miles.

Some large mouse-eared bats from the Tignahustes caverns of the Hautes-Pyrenees returned over a distance of one hundred and sixty-five miles.

Bats fly mainly at night so that visible landmarks must be

difficult to identify. Mice, with their low and limited vision would have an equal difficulty in recognising these but their keen olfactory sense may aid them over comparatively short distances.

Before we leave this subject of homing and migration I would like to relate very briefly the main facts associated with a few cases of homing I have been able to verify:-

The following case has already been mentioned in *The Mind of the Dog* (Country Life Ltd.), but as it is so unusual I will relate its main features:

From Professor H. G. Lamont, Belfast University

Dinah, a red setter bitch, in whelp, was sent twenty miles by train from Cookstown (N. Ireland) to Lurgan. Soon after her arrival she gave birth to five puppies. Within a few days she disappeared, puppies and all. She was found ten days later in her old bed at Cookstown, fast asleep. Her feet were raw and bleeding. Her five puppies were with her, busily sucking.

On her journey the bitch had swum the River Blackwater at Maghery Ferry, eighty yards wide and very deep, taking with her held in her mouth one puppy on a trip, leaving puppies on each side until she had got them all safely over and collected again.

From Mrs. Betty Smith, Broomhouse Farmhouse, Broomhouse Road, Edinburgh

A Border Collie was sold to a man living over a hundred miles away. The dog was driven ten miles to Kelso Station and completed the journey by train. He was met by his new owner and driven some miles to his new home.

He arrived back at his old home two days later having travelled cross-country across the hills through country entirely unknown to the dog.

From Mr. R. H. Hossack, Newholme, Slindon, Nr. Arundel, Sussex

A two-year-old Retriever dog was put aboard a destroyer, on her private trials, at Greenock and taken to Birkenhead. After a lapse of three weeks the dog arrived back home at Greenock. He had never previously been in this part of the country. He travelled by sea and returned by land a distance of two hundred and forty miles.

Cats have been known to travel just as far and return home. Generally speaking, however, a very small percentage of lost cats do return.

An Irish Water Spaniel disappeared from the country house of a friend of the author's. One morning, six weeks later, it was found dead in front of the house. Its pads were worn away and bleeding and the body was very emaciated. Probably this dog was stolen and carried a long distance by car, and eventually escaped.

My own Airedale dog, during the First World War, was taken by soldiers five miles by boat across a bay, then seven miles inland by motor lorry, to their camp at St. Anthony in Cornwall. The dog had never been in this part of the country previously. He returned forty-eight hours after his departure. He had travelled twenty-seven miles along the sea coast, swum the broad River Fal at King Harry Ferry, then twenty-seven miles back along the coast.

Sir Arthur Harris, Bart., Marshal of the Royal Air Force, recorded the following in *Country Life,* October 6th, 1960:

> "In 1923 an aircraft of the squadron I then commanded made a forced landing a few miles from Baguba, a village some 50 miles north of Baghdad. The areas between Baghdad and Baguba and Baguba and Hinaidi - our air base - can be described without exaggeration as howling desert.
>
> The aircraft remained undergoing repair for some ten days out in the desert. During that time there was no communication between the aircraft and the base at Hinaidi by land. During their stay in the desert the crew of the forced-landed aircraft befriended a yellow pi bitch which came out from Baguba with some Arab sightseers and decided to adopt the second pilot of the aircraft.
>
> The aircraft, after repairs were completed, flew on to Kurdistan and did not return to base at Hinaidi for some days. As it took off from the scene of the forced-landing the bitch lay watching its departure. The crew assumed that she would return to her owners - if any - in Baguba, and as she was in excellent fettle they had no qualms over leaving her.
>
> When the aircraft returned to base at Hanaidi the second

pilot found the bitch lying outside his bedroom door. She had either come some 60 miles across the desert direct, or along the dirt-track desert road 50 miles to Baghdad and thence 15 miles out to the Hinaidi base - where somehow she found the second pilot's room among those of some thousands of officers and men spread over an area of at least a square mile."

These are only a few of a great many examples we have collected concerning the peculiar senses of the dog. To attribute to the dog only an extra sixth sense, numerically speaking, might yet be a grievous understatement. Nor is it possible to assess the number of extra senses some other animals may possess - "extra" meaning in this instance senses additional to our own.

BIRD AND INSECT MIGRATION

As we now fully understand, the young bird setting forth on its first migration when only a few months old can have no knowledge concerning its destination. It must not be assumed that the bird flies deliberately of its own volition with a definite purpose in mind. It flies automatically and instinctively, and once again its complete plan of behaviour is in all probability conditioned by and set into activity by the secretion of a hormone into the bird's bloodstream. In birds of a migratory species during certain seasons of the year, in adult birds (usually following a moult), and at certain stages of adolescence in younger birds, a secretion from one of the endocrine glands (in all probability the thyroid although others may be involved), produces a feeling of restlessness, soon followed by a desire for company, an inclination to team up with other birds. This is why young swallows collect in groups on telegraph wires and flutter aimlessly about for twelve to thirty-six hours before the urge becomes so compelling that they set off on their travels.

If, *at any season of the year,* certain varieties of birds are injected with suitable and safe doses of extracts from the thyroid gland they will develop the migration complex and may set out on a flight even when conditions are entirely unsuitable. The pouring into the blood of small doses of the extract at intervals may also account for the *local* migrations of robins and other birds irrespective of long distance flights.

Caged birds similarly treated may become so restless that they damage themselves, or beat themselves to death in an effort to escape what has hitherto been acceptable captivity.

Many young birds, cuckoos and starlings for example, set out on their migration days or weeks before their parents are ready to fly, which makes it appear that the hormone output of the young bird fits in with a definite growth period, while in the older birds it is retarded by moult and the fact that ample food of the insect type is still freely available.

Birds can have no direct knowledge that an over-prolonged stay in this country in September, let us say, would be accompanied by a rapidly diminishing food supply, although migration dates are timed to avoid such a calamity. It is in the interests of the bird to wend its way by easy stages to greater warmth and a plentiful supply of food. It must not be considered that warmth is always an attraction since a good many birds, certain seabirds for example, some waders and even a few starlings, may even fly in a northerly direction.

Birds fly in all kinds of weather conditions, in darkness as well as in daylight, through cloud and dense fog. They avoid rain and storms whenever possible, but neither is an insuperable difficulty when need drives birds on.

None of the above conditions, in any way interferes with the bird's sense of direction, which implies that this is not dependent upon visual or other external conditions but upon some factor *within* the bird.

So far as vision enters into the picture it is believed that many birds on migration, and probably all when facing adverse weather conditions, close their third eyelids to cover their eyes, and "fly blind." It is true that in the centre of the third eyelid of many animals ranging from the polar bear down to the diving birds, there is a more or less transparent "window," but although in the small birds it may admit a certain amount of light, it is very doubtful if any birds, other than a few of the living birds, can actually see external objects through this "window."

Some of the explanations offered of this ability to follow a definite line of flight to a predetermined destination are as follows:-

(a) Terrestrial magnetism.
(b) Atmospheric electricity.
(c) Electromagnetism.
(d) Movement of the sun.
(e) The influence of ultra-short waves.
(f) The effect of wind currents.
(g) The search for warmer air currents.
(h) A developmental stage of the sex organs.

Not one of these has proved sufficient or even tenable in itself, nor does a combination of any of these alleged causes supply any answer to the problem.

Experiments with magnetism, electric forces, or short wave transmission have produced no results of statistical importance and so far as sex plays any part castrated birds migrate as constantly as those not subjected to surgical interference.

When we consider insects the picture changes. It is almost certain that most of the small insects are carried in rising air currents to a great height and wafted in the breezes sometimes for hundreds of miles before they return gently to earth. A great many are aphides which may migrate this way by the million. Traps set high over the North Sea have proved this windborne theory to be true in the case of many varieties of the smaller kinds.

The Report of the Rothamsted Experimental Station for 1951 dealt with two kinds of movement in insects. One, *purposeful migration,* occurred as in the case of the Large Cabbage White Butterfly, and two, a *passive drifting* on the wind might be involved, as in the aphidae (greenflies). Millions of butterflies crossed thirty to three hundred miles of sea without difficulty, the distance which separated England from France, Belgium, the Netherlands and Denmark. It was possible many had flown long distances overland before setting out across the sea.

It is interesting, and in the case of the White Cabbage Butterfly, gratifying, to hear that they were accompanied by a great many of the insects which parasitise the butterflies and lay their eggs within the bodies of the caterpillars.

The flight of birds has been followed a great deal of late years by ringing large numbers and liberating them. In this way the place of their origin and the time of arrival at the new location has been checked a great many times.

Quite a fair percentage of birds are sufficiently accurate in their direction-finding to be able to pair up with their mates of a previous season and make use of their old nesting site. House martins are particularly good at doing this and even if they were not mutually attracted for any other reason, the pair of birds which went through last year's matrimonial episode might be closely bound by a common territorial purpose in which they re-possessed themselves of what might well appear to them to be the finest nesting site in the world.

Young birds come into the world with the urge to migrate lying dormant, waiting for the time when their hormones will set the urge into practical activity. When the parent birds and their ancestors have all travelled in a south-easterly direction, let us suppose, the young birds will automatically select the same route, although why they do so is not clear. It is far more likely that the birds will reach their destination than that a young homing pigeon will arrive home from a fifty mile flight.

If these young birds were carried as soon as they could feed themselves, in a south-easterly direction by boat and car to the place where they were destined to spend their winter, it is likely that when their adult life was reached they would not return to the home from which they were dispatched but would fly even further to the southeast. This might mean, according to the part of the country, that they would end up in the sea, or in a country which could not provide their winter food.

Evidently a young bird experiences a sensation of comfort or assurance when it is travelling along the route followed by its ancestors, and again the analogy of the radio set and its varying signal strength when rotated, may enter into the picture. (See page 196.)

Perhaps the ability of a bird to head straight towards its

new home may bear some relationship to the ability which many people have to "divine" water or mineral ore, or sometimes even to locate a penny hidden beneath some part of a carpet. The two feats, that of bird and man, may seem very dissimiliar but when we compare them some of the seeming difference vanishes. The water diviner does not know exactly where the water or mineral lies if he is deprived of his divining rod. He has sufficient confidence to stand or fall by his results. The diviner, or "dowser," operates his rod by means of the muscles of his arms and body, and success in divining, or failure, depends entirely upon how he grasps his rod. He first points it at the ground, and grasps the two free ends one in each hand with thumbs uppermost. Holding the ends very tightly he flexes his elbows and then flexes his wrists, at the same time turning his folded thumbs outwards with the tightly clasped fingers uppermost. Drawing the two ends of the rod sightly apart he brings the point of the rod into the air in the erect position. This requires great concentration and strength. If he has held firmly to the two ends of the divining rod all this time, the tendency of the point of the rod will now be to turn down towards the ground. As the diviner advances over the territory to be examined he keeps the point of the rod uppermost by muscular contraction, a very tiring process. When the diviner stands over water or a mineral lode his muscles relax and the point of the stick relieved of tension, turns downwards and points to the ground.

In divining water the "dowser" has control of his muscles until he arrives at his destination (the water or the lode). At this moment his muscular control over the rod is lost. Standing over the water after the point of the rod has dropped the diviner cannot lift the point of the rod again until he has moved away from the site of his discovery.

It is just possible that the strength of the muscles of a bird's wings might similarly be influenced so that they operate normally only when approaching the prescribed area, and relax when the area is reached.

This suggestion may also be entirely without foundation,

but until we find the more simple explanation we can bear it in mind.

A feature of orientation is often observed, especially by younger people, when walking among trees, or entering a hall or a room somewhat overcrowded with furniture. To make the experiment stick a dozen bamboo sticks into the ground over a distance of twenty yards at random, so that to walk from the first stick to the last makes it necessary to pass ten other sticks with a choice of passing on their left side or their right side.

Unless one sets out with a determination to follow a set plan one will usually feel an urge to pass each stick on one side in preference to the other side, usually varying the side on which one passes different sticks.

The same thing happens when one rides across an expanse of ground through a herd of cattle. One feels an urge to pass round them not always to the right or left, but after the style of "polo-bending" to pass very frequently on the left of one animal and on the right of the next.

This is a sensation quite well-known to many people, and most persons under forty years of age will admit that on such occasions they do feel a directional urge. It is more pronounced in young persons approaching puberty.

This sensation, perhaps mixed-up a little with our own right or left-handedness, may be all that remains of our almost-departed sense of blind navigation. Perhaps we are at fault in wondering why birds and some animals retain so marked a sense of direction. Probably we should be asking ourselves why we have lost the ability to make use of our own.

Numerous observers have suggested that observation of the sun may help birds and prove itself a leading factor in directional flight. As successful migration takes place in darkness as well as in daylight the position of the sun as a guide can be at its best only an inconstant aid. Nor can observation of the North Star help a bird which has no idea why it is flying, or what the destination is that it seeks.

The same holds good for the theory that birds set their course by the sun when it sets, and maintain it through darkness, fog and low cloud, without ever going off course. As most day birds are completely blind between from an hour after sunset until sunrise, it follows that they must fly almost completely without the aid of vision by night, possibly, also, with the third eyelids closed over the eyes.

The part played by the earth's rotation and magnetism has been studied, as also has the possibility of a radar-like beam, or an echo-sounding device. None of these theories offers any promise after careful observation and experiment. In bats, their radar-like ability operates only over short distances.

We are not yet able to provide the complete solution of this problem of bird navigation.

We can assert with reasonable safety that migration behaviour is set into operation by the presence of a hormone in the bird's blood stream. This sensitises it to a more vivid appreciation of an instinct derived from its ancestors, which causes it to set out on a flight when the hormone exerts its optimum influence. It causes the bird to fly in the direction in which its ancestors flew at any particular season, and to keep on flying by easy stages when the distance is great, until the bird reaches the locality selected by its ancestors. How the bird knows in which direction its ancestors flew is not yet clear. It is possible that by some mechanism within it, the bird feels a sensation of discomfort when it goes off course which encourages it to correct the direction of its flight.

Time and Space

Many animals seem to have a time sense. This does not imply that they are able to tell the time or to divide the day up into definite sections. Cows know exactly when they will come in to be milked and will congregate at the field gate in readiness. This may of course be suggested to them by the fulness of their udders but all bovine animals whatever their sexual characters which are fed at regular hours exhibit the same knowledge of time, which may of course be associated with the feeling of hunger.

But dogs and birds also learn when it is worth while being

at a particular spot at a certain time and this is seen also in horses, pigs, sheep and poultry.

I think the best example of time and place consciousness was manifested by my retriever, Ben, and I have described his

behaviour in *The Mind of the Dog* (Country Life Ltd.) Briefly the story is that I purchased Ben from a gamekeeper on an estate seven miles from my home. I used to shoot every Monday and Thursday. Ben commenced to disappear from home on a Tuesday and Saturday, and would be away all day, coming home in the evening tired and dirty. Tuesday and Saturday happened to be the regular shooting days on the estate from which I bought the dog. When we watched Ben we discovered that he travelled the seven miles, there and back by 'bus, by courtesy of the conductor!

THE PSYCHIC ANIMAL

If we are unable to explain bird migration and homing flights satisfactorily we are even less able to account for some of the other manifestations of a peculiar sense which many animals appear to possess.

In my book *How Animals Talk* I have drawn attention to the case of a cat, presumed dead, which at intervals apparently returned in ghostly form to its old fireside. This cat was always recognised by the "new cat" although it was invisible to its owners, but it was seen in normal fashion and commented on by a complete stranger.

I would like to mention a few other authenticated examples which will explain better than many words the type of behaviour to which I refer.

In Chapter One I referred to the dog which knew when its master was on his way home by train. Here is another somewhat similar case recorded by Mrs. K. C. Heffer, Pound Close, Twyning, Glos.

Mrs. Heffer had owned an Elkhound, of very placid disposition, since it was nine months old. At the time she was living in Cambridge and made frequent trips to London, but never knew when or at what time she would return. On her arrival home she was always surprised to find the meal hot and waiting and the tea brewing in the pot. When she inquired how it was known that she was returning at that hour, her maid informed her that whenever her train arrived at the station, which was four miles distant, her elkhound would jump up, wag her tail and run to the front door, where she would sit anxiously waiting for the twenty minutes which the 'bus took from the station to the house. As trains are entering the same platform at all times of the day such behaviour can hardly be accounted for by simple explanation.

A great many people have encountered animals which fight shy of certain houses or gardens.

I, myself owned three terriers, one a smooth fox terrier bitch and the other two Jack Russell terriers. None of these, separately or together, could be persuaded even on the pretext of a ratting expedition, to enter the overgrown and dilapidated garden attached to the ruins of an old house, which was (as most old houses in Cornwall are) reputed to harbour a "ghost."

One explanation offered was that they feared snakes, but although it is not unlikely that the garden may have contained some vipers, I always found it difficult when walking with these three dogs on the cliffs, to prevent them chasing a snake when they saw one. I never discovered the real reason. A friend who owned a rough terrier of Airedale type discovered that his dog, too, had an equally strong objection to entering the same garden.

There was a certain corner on a bye-road near Falmouth which very few horses (in the days when horse traffic was the usual method of transport) would pass without shying, or coming to a dead stop. There was nothing unusual in its appearance and the corner has now been dispensed with but nothing was ever found to account for this exhibition of equine dread.

Mrs. Betty Smith of Broomhouse Farmhouse, Edinburgh, when farming on the Borders, had in her employ an old shepherd who owned a Border Collie. The shepherd used to depart with the show sheep to the various exhibitions. The Royal meant an absence of about ten days, the Highland about six, and so on. When this shepherd left he followed his own inclinations and no person knew to a day or two when he might return. The shepherd lived a mile-and-a-half up the valley from the farm where he was to unload the sheep. Whenever he arrived at his home Bob, the collie, would come running up to meet him to drive the sheep home.

Over a period of years Bob was never absent, never late!

In November 1947, Mrs. Smith's husband left by car to go to Melrose, a distance of thirty miles, to bring home his mother and his son. They were expected home in time for tea.

At 4:20 p.m. her three bulldogs and an elderly bull-mastiff were asleep on a mat in front of the fire when all four jumped up and rushed madly to the front door, as though they had heard the car arriving.

Later, Mrs. Smith was to learn that the car crashed on ice on the Soutra Pass. The granny was killed outright and her husband and son seriously injured.

The accident occurred at 4:20 p.m. precisely.

In company with many other folk, I could recount many stories of dogs, which sometimes by day and sometimes by night, have stood trembling, with hair of neck and back standing up on end, while growling fiercely all the time they watched some object, invisible to human beings.

Whether the dog, like some people, suffers from hallucinations I cannot say. It is certain however, that on at least three occasions, such hallucinations must have affected two dogs simultaneously.

Chapter V

SOCIAL INSTINCTS

Perhaps, to be quite up-to-date I should have avoided the above heading and substituted "Social appetites," or "Social drives," or some equally modern term. But as in the long run all of these mean pretty much the same thing, I propose to retain the original expression. An instinct is inherited and some students of psychology are not at all certain that behaviour patterns *are* inherited. This view introduces fresh difficulties.

A cat catches mice and tortures them in precisely the same way as every other cat, and we could presume that the behaviour was instinctive. Yet, if a kitten is reared by hand it does not always catch mice, nor kill them. It has not learned the art from its mother.

The word "social" might come in for criticisms since it does not apply so strictly to the lower orders of animals as to the higher.

Animals frequently congregate for reasons sexual rather than social, or are together on account of familiar reasons, neither of which would be considered by a great many people as being strictly social.

Flocks of birds come in the latter category but in these again, congregation is directed towards mass protection and communal feeding, and although birds like being accompanied by others, each remains an individual without social ties of any kind.

There are no clubs and no afternoon tea parties in the animal world, and everything is very matter of fact with an eye to survival rather than companionship.

If we wish to be very analytical we may even query whether the congregation of human beings is always, or even often, organised for purely social reasons, and whether the majority of human contacts are prolonged beyond the limits of bare necessity.

Very few animals form social ties in which sex or appetite does not play a part.

Men and women have an instinctive fear of being left entirely alone in life. They dread old age and loneliness, and they make provision against the latter whenever it is possible to do so.

In the wild state age is not kind, and the old animal is far safer alone than in company.

In the animal world the law of *meum* and *teum* has no practical application since every animal regards everything within its reach as its own.

Man has had it drilled into his mind from the time he first began to understand human speech, that almost everything he sees belongs to someone else, and that anything he sees and wants can be obtained only on terms. These usually are expressed as cash, and as the possession of this desirable commodity entails the exercise of mental or physical labour - or both - one soon learns to calculate whether possession is worth the price it demands.

Apparently the principles embodying barter are beginning to be understood in the animal world. It is reported that apes and some of the more intelligent monkeys are being given jobs of work, and for their successful accomplishment these animals are being rewarded with coins, or money tokens, which they insert in a slot machine in exchange for food or sweetmeats. Animals learn in this way to carry out a number of mechanical feats which they would never contemplate doing unless they had been taught that cleverness brought its own reward. According to the London *Daily Mail* of November 12th, 1960, a Texas furniture factory owner has three chimpanzees working on an assembly line making rubber foam sofas. They are paid with bananas.

So far as can be judged from the eagerness and from the emotional expression shown by these apes and monkeys, they are decidedly of the opinion that a little work, well rewarded, makes a welcome addition to the daily round. If this sort of

experiment is continued it may soon be discovered that animals possess far greater intelligence than they have been credited with, and there is no knowing where the system may end. It may be that when some ape, a little ahead of the others, starts making a book, and all the remaining apes discover that there is an easier way of making money than working for it, the experiment may end in disaster.

I owned a Rhesus monkey nearly forty years ago and I think that he may take credit for being one of the first monkeys to introduce the system of working for money. Peter's conception of what constituted work was not unlike that held today in far higher circles than he patronised, although it was one I could never encourage. Moreover, Peter may well have been a miser at heart and he would never have dreamt of spending his ill-gotten gains on sweets. I suppose if he had done so his business would have gone into liquidation because Peter really depended for his income upon the number of people who *brought sweets to him.* This was his system:-

Peter spent most of his days in a state of freedom. He had been neutered when a baby and had no vicious propensities whatever; in fact he was a very likeable individual, extremely fond of human society. His one drawback was that unknown to most of his acquaintances he was a most expert pickpocket, and under the guise of cuddling up to any stranger who brought him sweets he would quietly divest him (or her) of everything portable he could surreptitiously transfer to his cheek pouches.

He would collect a lady's ear-rings, her necklace and even her wrist watch on occasion, though never the contents of her purse, and he would stuff them all into his pouches before the lady had any knowledge of her loss.

He was very honest regarding money; he collected only jewellery and trinkets, especially engagement rings, which he was prepared, even eager, to hand back to their owner in exchange for bright new halfpennies fresh from the mint. I am quite convinced that jewellery meant nothing to Peter; his interest lay in sterling.

His procedure was very simple. When Peter's pouches, the roomy receptacles within his throat, contained a pair of valuable pearl ear-rings, let us say, quietly removed while their owner was cuddling him and feeding him with chocolates, he and I would resort to gentlemanly bargaining.

Firstly, I was permitted to palpate Peter's pharynx, to make sure that he was being perfectly sincere and that he really possessed the lost property. I would bring out from my pocket a bright, new halfpenny. I kept a large stock of these in readiness. Peter's next move was to dive one hand down into his throat and produce one ear-ring. In return he could be permitted to hold one half of the halfpenny while I held the other half. I had found it unwise to part with any money until the whole transaction had been finished as sometimes he might decide to postpone the completion of the deal. I used then to exhibit a second halfpenny and down would go Peter's hand and up would come the second ear-ring.

But that was not all, as Peter frequently indicated. On production of my third halfpenny he would make another dive and bring forth the stump of pencil he had extracted from my own pocket the previous evening. Simultaneously, and very seriously, the exchange would be completed.

I had then to proceed to the nearest tap, wash and dry the ear-rings and return them to their anxious owner intact. Honour was satisfied.

Peter would fetch his bright golden syrup tin (empty of syrup fortunately) and drop into this the halfpennies to join a lot of others; then he would replace the lid and carefully balance the tin on the bar of his swing. When, at night, I went to put Peter to bed he would be sitting up waiting for me, with the syrup tin open on his knees letting the bright coins flow through his fingers in a way which would have gladdened the heart of a Shylock.

On only one occasion do I remember him ever damaging any article he borrowed in this way. On this he snatched the pince-nez off the nose of an elderly gentleman who had drawn his car alongside of ours in a traffic jam. Before we could enter into negotiations the cars moved on and we lost contact. In annoyance I imagine, Peter chewed the pince-nez, glass and all, into small pieces and swallowed the lot.

When we consider what lies behind such behaviour, we must admit that the whole pattern, devised by a monkey entirely without prompting, suggests some ability to reason. Undoubtedly the instinct to purloin was instinctive, and it is one which appears to be present in all the higher Primates. In the human species it is controlled by social and legal regulations, while every child is taught not to covet that which belongs to a neighbour. But it was Peter who first made it clear the articles could be redeemed for a consideration.

The fact that the consideration should take the form of bright halfpennies is interesting. It should be mentioned that Peter commenced collecting halfpennies some time before he started playing Robin Hood with lady's ear-rings, and it was only after a deal of experimenting with bribes of various kinds that I discovered that bright new halfpennies meant more to Peter than a diamond engagement ring.

The question then arises - why? The halfpennies were never converted into edibles, as happened in the case of the apes who did legitimate work and popped the coins they received for their labour into a slot machine in exchange for sweetmeats.

Did Peter value these coins because they were bright and yellow, from a purely aesthetic reason - or because to him they represented personal possessions, cherished the more because he had devised the means by which he obtained them?

Or, are the instincts of the miser present in all the higher Primates?

Do we all wish to hoard for hoarding's sake - and is it because we find it economically impossible to do so that more of us do not become misers?

Another point arising from the consideration of Peter's misdemeanours concerns the aspect of guilt and the matter of conscience. I have owned a good many monkeys and an ape or two, and I realized that it was natural for them to steal. But if another monkey had stolen my penknife, let us say, he would have sat in a sullen mood, watching me closely to see if I suspected him. When he realised that I did so he would bolt, scream at me, attempt to bite me if I approached, and as happens always with an enraged monkey he would have exhibited symptoms of violent diarrhoea, which would have assisted him in making his escape from the house.

But Peter never reacted this way. As soon as he had pouched the jewellery he would endeavour to attract my attention by coming to me and handling his pouches from the outside. Then if I did not appear sufficiently intelligent to grasp his meaning he would introduce finger and thumb, draw the article out until it protruded between his lips then swallow it quickly as soon as he knew that I had seen it.

On several occasions when I appeared particularly dense, he had withdrawn a ring, showed it to me in his hand, then quickly popped it back again into his mouth - a clear invitation to me to commence bartering for its return.

He would sense my growing concern and deliberately trade upon it. A single halfpenny would serve for a small article such as a diamond engagement ring, or a pearl ear-ring, but a necklace was an entirely different matter and one bought it with halfpennies, inch by inch, as Peter extracted it a little at a time.

The worse bargain I made with him was for a very cheap imitation pearl necklace which came unthreaded in his pouches. The pearls came forth for one halfpenny a time, and before we had recovered half the necklace I had to postpone the deal until I had been to the bank for another pound's worth of halfpennies.

Peter's valuation was essentially simian in outlook. The more there was of an article the more valuable he considered it. I could only console myself with the thought that on one occasion I had effected the return of a diamond solitaire ring, worth a small fortune, for the outlay of one halfpenny.

In abstracting these articles from the person or pockets of his human acquaintances Peter must not be considered to have acted with dishonest intent. Animals have no understanding of what we term right and wrong, and as I remarked previously the whole world belongs to the animal and there is no such thing as individual ownership by right. It may be difficult perhaps for one dog to take away a bone from a larger dog but if he can devise a means of so doing the small dog is acting entirely within his rights. The right to be in sole possession of an article is entirely a human invention.

I have no doubt some of my readers will be inclined to quote examples of animal behaviour in an endeavour to prove that some animals practise deceit and treachery, which should be considered evidence of wickedness. They will bring up simple examples, the spider with its web or its trapdoor, the angler fish with its "rod and bait," or even the havoc wreaked by a fox in the hen house.

Human criminals plan their nefarious designs in their brains, their crimes are premeditated, and sometimes carefully worked out in detail beforehand. The criminal is fully conscious that what he proposes to do is wrong, and that it will cause suffering and hardship to other human beings. In the animal these "sins" are not designed in the brain but are already contained within the mind when the animal is born, and they form a part of its normal pattern of behaviour. The fox takes no delight in torturing its victim and although it may commence to devour the chicken while it still lives and feels, the fox has no knowledge that it is causing agony, or even that other animals feel pain.

Animals possess only simple brains and are quite unable to produce a human response to an animal stimulus. They are incapable of deceit. When animals play a part in furthering human wickedness it is because they have been taught by their human companions to behave in a particular way.

CONGREGATION

Animals congregate for a variety of reasons.

Men do so in order that they may live the one upon the other, and no wise man shuns the company of his fellows since trade and barter are his means of existence.

Man can live in a community only when a considerable number of individuals are grouped within easy reach of one another, and as a species man is not adapted to leading a solitary existence. Not only does he need companionship but he has a fervent desire to assert himself, to express himself as an individual, and to be recognised by his fellows as being somebody of importance. These are natural instincts inherent in the race. As an individual, too, he may be intolerably lonely in the heart of a big city, but if he wished to withdraw himself from society he would find it impossible to live by what society members considers honest means.

Many cherish the idea that they would like to be self-dependent, shake themselves clear of the fetters which bind them to their fellow-men, and stride forth alone into a world of glorious freedom.

But only one of the lesser animals can do this successfully. Human society depends for its existence upon the cohesion of its separate particles, a linking together of millions of units. No man can remain an individual yet detach himself from the shackles which bind him to society, if he possesses a normal human mind.

But there are countless human beings today - probably there

have always been - who possess minds little more advanced than those of the lower animals, and such individuals are called upon to accept human responsibilities, and produce the responses demanded by civilised society.

In the wild state animals live their whole lives in competition, firstly, between species and species, and secondly, between separate individuals forming a part of the same group. Their first object, especially so far as species competition is concerned is food, and their ambition as individuals is to eat at one time, without interference, as much of the food available as can be swallowed.

This applies particularly to the ruminants, and to grazing animals generally, which require a long feeding time to overcome the slow rate of assimilation of wild herbage in process of digestion.

Animals in general have longer periods of abstention from food and drink, at times, than men usually experience, and their rule is to eat at any hour of the day or night when opportunity arises. Men eat not because they are hungry but because the hour has come when society ordains that they shall eat. The fact that men and women are employed by other men and women for set hours in a day, makes social eating times imperative. It would be a hopeless state of affairs if employees knocked off at all odd times in order to eat.

Animals in freedom have their territories, while man has his sales shops, his workshops, farms and factories, in addition to his own personal territory, his home. In his working territory he may be producing food, clothing or appliances for himself and his household, but more particularly for other human beings and their homes. The money he receives is expended not only on food and clothing and a house with a roof, but also upon a great variety of articles, the use of which a domesticated animal will not even have any understanding. Probably a horse knows the use made of a carriage or a cart, and of harness, but no dog understands the purpose of a refrigerator, or a washing machine, nor would one appreciate the use made of a billiard table.

Man is the only animal who has advanced to the stage when

he can by his own energy earn sufficient to keep the whole of his family fat, in idleness. In modern times the number willing to behave in this way is rapidly diminishing. Nowadays, the whole family earns, and the wife can bring home sufficient to enable her to employ some less intelligent woman to rear her children for her.

Man has attempted to tame and domesticate animals ever since the days when the giant sloths, nearly the size of an elephant were still in existence. There is evidence from fossils that man kept these sloths in captivity and also that he attempted to tame the primitive cave bear. Why he should do so is a little obscure, but it goes to show that the desire to keep animals in domesticity has been handed down to us through countless generations.

Dogs were probably domesticated in the latter part of the early Stone Age, or the late Stone Age, from the eighteenth to the sixth millenary B.C., but cats which were solitary animals, probably did not form any close links with mankind until about four thousand years ago. It is quite likely that wolves, captured as cubs, were reared by man even before dogs assumed their special characteristics and became a distinct breed. As I wrote earlier it is likely that it was the brachycephalic "sports" of the dog family which first developed into "pets," as opposed to dogs of the wolf type used for hunting, guarding, and in all probability as fighting allies against other men. Horses were used by mankind from about six thousand years B.C.

It does not follow that primitive man maintained even the brachycephalic dogs for pure love of the species. No doubt children all those years ago had similar instincts to those exhibited by children of the present day, and a warm, cuddly puppy was not only pleasant company by day but especially so on cold nights before hot-water bottles were invented. It is also likely that when the puppy attained a convenient degree of fatness it was disposed of in the same way as farm wives dispose of the "cade lambs" they have reared on the bottle. Roast puppy, even today, is still a delicacy in many lands.

The process of domestication involved the discovery by the

person who was animal conscious, and willing to spend time and trouble to help bring this about, of an animal willing to become domesticated. Not every race of human beings possesses this capacity. The African natives living all their time among a great variety of wild animals show no natural affection for them, and very little sympathy. Their sole object is to catch and kill, and since they have made closer contact with the white races and have discovered a market for skins, horns, ivory and rhinoceros horn, the fate of the living creatures of this vast continent lies precariously in the balance.

My own opinion is that almost any animal of manageable proportions can be domesticated. I would fight shy of the African elephant which is temperamentally unknowable, and such animals as giraffes, which apart from lack of head room in most of our premises, do not live in domesticity with any relish, and like a few other animals cannot tolerate any form of restriction. But even the most unusual creatures respond to kindness, gentle handling and comfortable living conditions, until, as I mentioned earlier, the breeding urge disturbs them emotionally and temperamentally, and they may become what the layman terms "vicious."

The emasculation of animals *at an early age* keeps them docile with very little difference in their general behaviour or appearance, and monkeys, for example, which otherwise become spiteful and untrustworthy at four or five years of age, will live up to ten or twelve years in happy domesticity. To learn a great deal about an animal one needs to study it in its adult life quite as much as during adolescence, and emasculation of the animal makes this possible. It also does away with the frustration and unhappiness experienced by the unusual pet which has little prospect of being provided with a mate.

One of the reasons for which animals congregate may be the survival value associated with group protection, but this is not always the reason.

A collection of limpets on a rock may show little increased aptitude for defence, nor can they be considered creatures likely to be socially inclined. When animals are as limited in their field of travel as limpets it is necessary to have individuals within easy reach of one another if fertilisation is to be effective in maintaining the numbers within the species.

Again, when a thousand ladybirds lie dormant on a rock there is a greater chance of the majority escaping the notice of their enemies than if they are spread about in groups of twos or threes over a wider area.

Even among herds of large animals such as deer, it is more economical to lose one or two lone sentries from the outskirts than to sustain an invasion of the whole flock by wolves or other predatory animals. There is always a superabundance of males which are more easily spared than the valuable females.

Small birds gathered into large flocks are far safer than when they lead solitary lives. The sudden uprising of a flock produces a "*confusion effect*" upon a falcon or a hawk, which seldom makes a capture unless birds miss their timing and become detached from the main body.

The territory in which an animal lives usually provides it with the food it requires. If it happens to be a vegetarian and is hard put to it to find enough to eat, it may make inroads upon human territory, or upon that of others, since in the animal world all possessions are shared. But the owner of that particular territory may not agree with this doctrine and so battle wages.

If the animal is carnivorous it may be no respecter of territorial immunity. It may belong to a species such as the lion family, which kills one animal to satisfy its hunger, or it may be a more rapacious animal such as a fox, which kills far more hens in a run than it can carry away. The fox in higher circles would have qualified as a sportsman, and would have delighted in the mass murder of game.

The human sportsman breeds and rears his animals in order that he may enjoy a Roman holiday. The wholesale slaughter of defenceless creatures appeals to the sportsman's aesthetic tastes since he perceives something of beauty in the destruction of animals which have no means of retaliation.

But even the kindest-hearted men may feed animals and make them comfortable and fat in their contentment, since humanity has carnivorous tastes; and we who relish our steaks, our pork chops and our poultry, must accept an equal share of the blame, if there be any.

Fortunately for all of us cannibalism is out of fashion. Similarly, lion does not eat lion, though dog may eat dog, and wolf eat wolf. The father wolf, good husband as he is, has a weakness for wolf cub, but the mother wolf will guard her young against him with her life, being fully aware of his weakness. One does not hear of fox eating fox, or bear eating bear, though in times of stress most animals will stretch a point if they are both carnivorously-minded and hungry. Foxes will eat berries and roots when they cannot get flesh, while apes and monkeys catch and eat birds and mice when unable to get fruit and nuts.

I am reminded of a sad incident which came to my notice when I was a college student a great many years ago.

A certain poacher caught a stray, starving cat in one of his gin traps. He owned four ferrets with which he earned his living and before retiring to his bed he threw the injured cat into the ferrets, for their supper.

In the morning he found only one animal alive — It was the cat!

On occasion there appears to exist between different varieties of animals a kind of symbiotic relationship, or at other times one animal may be parasitic upon another. Quite frequently symbiosis is directed against parasitism, at least as regards the ecto-parasites which live outside the animal body. One commonly sees starlings and jackdaws perched upon the backs of sheep or cattle, busily ridding them of lice, keds, or ticks, at the same time procuring a meal for themselves. Birds sit on the heads of a hippopotamus, or on the back of a rhinoceros for very similar reasons, and even the crocodile has its companions.

In the human race symbiosis exists but on a rather higher level than among the other animals. It is the ordinary thing for one man to make a pie for another man to eat, who may make the clothes the pieman wears. In fact, all human beings live one upon the other and by the good graces of one another.

We have to distinguish between commensalism and symbiosis. In symbiosis both partners benefit, while in commensalism, which means "feeding together," only one partner really benefits.

One surprising example of commensalism is instanced by a South American parakeet which builds its nest on a termite's hill. The termites, large and ferocious ants, swarm out to repair the damage but for some strange reason do not attack the parakeets or the nestlings, although they would devour any other living thing. The parakeets gain because no man or animal will dare interfere with the nest or eggs, but why the termites allow the parakeet to stay, or what, if anything, the termites get out of this bargain has never been discovered.

Very few animals are fond of complete solitude - the number of individualists is less than in the human race. Birds which live in flocks, will prefer to pair off during the breeding season, but those which migrate will come together when their thyroid glands become increasingly active after the brooding period is over. They may congregate for purposes of feeding, perhaps to make a communal raid upon gardens or protected feeding grounds, which they would hesitate to invade singly; or it may be the existence of food supplies which brings the solitary birds into congregation. Other birds may feed in small numbers, or singly by day, and come together by night in a common roosting place.

Bats may congregate during winter for purposes of hibernation, and the question has lately arisen whether some birds c.f. nightjars, may not spend some part of the winter in hibernation in some countries. Snakes will often hibernate in large numbers all together in one hole in the ground, attracted there by the scent of the others.

Most of the animals which live in communities, whether

herds or flocks, are made up of a number of family groups, together with individuals in no way related to these groups.

Most herds and flocks make their commencement entirely by in-breeding, which may continue through a good many generations. In time a few individual males from outside will become attracted and may make their influence felt, or they may be driven out again by the existing males.

In-breeding seldom appears to produce any ill-effects among wild animals, perhaps because those which survive a wild life carry few defects likely to prove lethal. In-breeding carried out to the same extent in domesticated animals such as the man-made breeds of dogs, might produce quite different results both physically and temperamentally.

It is not always easy to determine how flocks or herds originate. Some accept outsiders without question if they care to attach themselves, but the majority of communal groups depend largely upon breeding and retaining their progeny in accordance with the amount of foodstuff available for their upkeep.

The Elm Tree gang of sparrows, which have their headquarters in the tree across the road from my study window, numbered sixteen members during 1957 and again in 1958. All intruders were driven off, and it appeared that sixteen was the maximum number the district would accommodate. But when I guaranteed their rations it soon grew larger. In 1959 it contained twenty birds, and in 1960 thirty. All these extra birds were chicks of existing hens who brought their broods into my kitchen for inspection - and extra tit-bits.

There are still two cockbirds to each hen but monogamy goes for nothing in the sparrow world and each hen regards herself apparently as the legal wife of one particular cockbird, and mistress to the remainder. If I ceased to feed these sparrows their numbers would soon fall to the original sixteen. Quite a lot would desert, and perhaps start another gang which would seek an area where food was more plentiful. Young sparrows, when underfed, leave the nest before they are fully fledged and fall victims to the cat, who also has to live.

Young males introduce themselves into a flock or herd more

easily during the anoestrous, or non-breeding period, when the resident males are practically neuters and far less pugnacious. They usually hang around the fringes of the herd, act as sentries and are often carried off by a predatory animal. If in this way they save the life of a female they are doing all that is required of them. If the females can be saved the number of males required to maintain a normal rate of herd fertility need not be very large.

I do not propose in this book to discuss the fighting which goes on in flocks and herds between the males, other than to say that while an occasional fight is serious, the majority of combats are mainly play-acting between two participants, each of which is awaiting a reasonable excuse to quit without loss of face, rather than seriously attempting to kill his adversary.

The matter of social status is apt to become a little overshadowed during the mating season, and under all the bravado and make-believe in fighting, the youthful male, even when more active than the aged animal, still gives way unconsciously to age and territorial authority.

Influence of Suggestion

When animals congregate they tend to follow a similar pattern of behaviour under the influence of suggestion. By suggestion I mean the tendency to develop ideas by noting the behaviour of other members of the flock or herd. This reminds one of the behaviour of children when one cries "Wolf," or of a cinema full of people watching a horror film.

Deer are much nearer to being natural animals than cattle. If you watch a herd of these you will observe that they are very highly strung and react very rapidly, and also very frequently, to stimuli of such a mild nature that they would be ignored by domestic cattle.

It is seldom that all the members of the herd feed at the same time; usually two or three or more are moving about slowly with heads raised high, ready to give the alarm signal on the least provocation. Occasionally one animal will become tense, run a step or two, then stop and look around to see what effect such behaviour has upon the others. If they show no interest the confidence of the nervous animal is restored and quite possibly it will lower its head and commence to feed.

At times, the pretence of fear spreads rapidly, one or two others may adopt the short run tactics, and it is likely that the whole herd will decide that some unrest is afoot, and they will move to a fresh feeding ground.

If something does really disturb one of the herd, whether it be a sentry on the outskirts, or an animal within the centre of the herd, its abdominal muscles will suddenly contract in response, as also will its diaphragm. Such contraction forces air upwards through the larynx and produces a low grunt. This is immediately picked up by the ears of all the members of the herd who will at once move off at a fast pace. The sound may be less in volume than a human "hiccup" but it is heard or sensed in some way by animals as far away as fifty yards from the source of the sound.

In a different way the impulse to sleep, to eat, drink, walk about, take short hops or flights, to stage a fight, or indulge in some form of exhibitionism, can spread rapidly through a flock or herd, purely as a result of suggestion arising from some impression transmitted by another member. Exhibitionism, hopping and running after imaginary insects, as performed by birds, especially by waders on a sandbank, may represent some aspect of play, but it is incited by this form of suggestion and may influence a number of individuals at one time. The term employed to describe this tendency to follow the antics of another animal in response to a visual stimulus, is *sympathetic induction,* or *social facilitation.*

Mobbing

Communal attacks are sometimes made by animals upon one or more individuals of either the same or a different species. Birds, cattle, elephants, rhinoceroses (more often singly) and carnivorous creatures such as dogs and wolves, frequently make such attacks.

In the birds the mobbing of an owl or a hawk during the

hours of daylight is a common occurrence though in a great many instances the individual least disturbed by the whole procedure appears to be the owl. The hawk flies away with wagtails or sparrows in pursuit, but apart from the fact that it interferes with the hawk's hunting programme little damage results. The fact that the small birds keep together and show demoniacal fury probably impresses the bird of prey, faced by an unusual situation which it may not fully understand. In any case neither of these birds usually dives into a flock but selects a straggler on its outskirts, or a lone victim.

Rooks are notorious for their congregations and in many cases this word describes their meetings well, since there is a saying in the Westcountry, "Rooks go to prayer meeting each Sabbath morn."

This is attributed to the fact that on a fine Sunday morning when everything was quiet, as it so seldom is nowadays, large flocks of rooks would settle in a field and in the surrounding trees. Usually a bird perched high on a branch would utter a long and loud succession of "caws," which the local farmers declared was a sermon, preparatory to demanding a sacrifice.

It is a fact that after a noisy but otherwise peaceful meeting, the rooks would frequently make a concerted rush upon one solitary member of the flock and proceed to peck it to death. Whether this represented mass hysteria, and if the victim presented an unusual impression through being ill or injured previously, has never been plainly demonstrated. It is a fact that most animals have no mercy on a casualty, and this may be an explanation of the rooks' behaviour.

Animals see others as members of a particular group, and from their experience they are aware of the specific type of behaviour common to that group. If animals exhibit any visible deviation from the normal they become common enemies - creatures under suspicion - and the object of the whole of the animal kingdom is to destroy them.

It is unfortunate if a blackbird carries two or three white feathers on its head or in its wings, or even if it has a damaged leg or foot. Individual birds will give it a wide berth while a flock may set upon it and peck it to death. An all-white blackbird, not an uncommon variant, rarely finds a mate, and may be chased on sight by birds all and sundry.

If the "boss-cow" of a herd becomes lame or debilitated, she will lose her position and become a target for every cow to aim at. She will be hustled and bustled, pushed into corners and generally ill-treated by even the lowest member of the party.

In the same way a month-old chicken which has its feathers soiled or wet, or has a wound on its head or body, will in all probability be pecked to death by all the other chickens in the coop. One of the greatest difficulties found in rearing broiler chickens in closed communities, is the habit of feather picking, which quickly leads to wounds, and these act as a stimulus to every chicken to set about the damaged bird and put it to death. As soon as this has been accomplished the murderers begin to look among themselves for the next victim. This type of behaviour is engendered by boredom.

Animals recognise only that which is normal and fits in with their accepted picture of behaviour pattern. As I have mentioned previously a dead dog lying in the road means nothing more than an obstacle to be avoided. Live dogs walk about and bark; something which does not walk about and bark cannot be a dog. There are live dogs and dead dogs, but a live dog knows nothing of death.

A large dog, reared with a child, will probably learn to protect it, and in a spirit of comradeship will accept a deal of rough usage with good temper, and will refrain from biting, even when mauled about and hurt. The child is acting in the way the dog expects of it and the pattern of the child's behaviour becomes fixed in the dog's mind.

If that same child were to run along in front of the dog and suddenly fall in a convulsion, or sustain some injury so that it struggled on the ground and was unable to rise immediately, the dog would develop an anxiety complex. It would realise that some form of response was required, but as this was an entirely unforeseen circumstance it would be divided in its mind as to the nature of the response.

A child struggling on the ground, or a child unconscious, bears very little resemblance to the picture of the child already fixed in the dog's mind. In addition, the behaviour pattern no longer corresponds with one the dog has learned to accept.

The child has therefore vanished, and in its place lies a struggling, or inert form as the case may be, which conveys no meaning to the dog.

Two courses lie open to it - it may run away and ignore the recumbent form, or it may do as the chicken does in a similar circumstance, it may attack it. A bitch will almost invariably adopt the former course and ignore the whole situation; a masculine animal is more likely to attack, not the child it loves, but some remarkable object which is no longer a comprehensible figure; one which has taken the place of the child.

The above refers to the individual dog but a pack of hounds will invariably attack an object presented before them which they do not understand, in somewhat similar circumstances. In *The Mind of the Dog* I have told the story of the hunt terrier, a friend of the whole pack. On one occasion when some mild fighting took place in the kennel, the kennelman whipped the leading dog. This dog then bit the dog in next social position, who attempted to bite the next dog down the scale. Before he could do this the whole pack, thoroughly disorganised and at a loss as to how to respond to this remarkable situation, set upon the hunt terrier and tore it limb from limb.

In the circus the lion-tamer controls his charges not by superior strength but by being able to maintain his number one position, mainly by anticipating every move and every intention of each animal before it can be set into operation. He relies largely upon the escape reaction, and on making use of the escape distance peculiar to each separate animal, which the tamer knows to an inch.

But every lion-tamer knows when he enters the cage, especially when the animals are of mixed species, such as lions and tigers together, that one momentary loss of face, a stumble, or a fall, will instantly break the spell, and that every animal in the cage will at once set upon him with full intention to complete his destruction.

Exactly the same applies to a bull led into the show ring. The bull is more powerful than seven men but while it is under the influence of man's mental superiority it may behave in an orthodox manner. If the man leading the bull is pulled off his feet, so that he falls in front of the animal, his chances of escaping uninjured are very small.

A great many varieties of animals respond to certain inflexions of the human voice. The vicious horse used to be controlled by the sibilant hiss which emerged between the groom's incisor teeth as he passed his brush below the animal's belly and flanks.

In my young days I was told many stories concerning "The Whisperer," a man who came to England, possessed of the power of calming the most dangerous Thoroughbred stallion by means of a whisper.

The story goes that he was the subject of a very large bet as to whether he would tame a particular man-killing stallion, or whether he would be found dead when the gates into the animal's yard were again opened.

Actually, the stallion was discovered lying on the ground, more or less restrained by "the whisperer's" own particular appliances, while the gentleman in question reclined comfortably on the horse's neck.

What the end of "the Whisperer" was I never learned, but I have discovered from much experience with horses and bulls, that sooner or later the human being usually makes a mistake, often a fatal one.

THE OLFACTORY SENSE IN PACKS, FLOCKS AND HERDS

Many animals are provided with glands of various kinds which open onto the skin surface and secrete a scented material which identifies the animal, so that it is recognised by every other of the same species.

The simplest example is the dog. Beneath its tail it has two sets of scent glands of different sorts. On either side of the anus is a gland known as the anal gland which secretes a little fluid, with a distinctive odour, on each occasion the dog passes a motion. The second set of glands consists of a number of very small perianal glands, which open onto the skin surrounding the anus through a series of pores. These glands are sex indicators and a dog or bitch can recognise the sex of a strange dog by the odour they provide.

When two dogs, unknown to one another, meet they do not go into head-on fray but charge past one another, their sides almost touching. As they pass each notes the smell of the other's posterior.

Neither dogs nor bitches distinguish the sexes by visual means, even when the signs are plain to see, but they make no mistake when they trust their noses.

One hardly need mention the skunk which uses its anal glands for threat and defence. Bad as its reputation may be it is not alone, except perhaps in its actual technique in which it has no competitors in quality, or aiming ability. The badger, too, possesses a pair of anal glands which secrete a fluid smelling like musk. They expel this when the animal is alarmed.

Other animals with olfactory armament include the polecat, stoat and ferret, the dog fox, and among birds the hoopoe. The female hoopoe, aided by her young ones, can eject an evil-smelling fluid, which is nearly as efficient in giving a permanent odour to one's clothing as that provided by the skunk. Some of the bats can eject an unpleasant-smelling fluid from a gland on either side of the face. If one handles an oppossum, especially one of the striped variety, which is 'shamming dead,' one not only gets bitten hands but one's apparel will also be unwearable until it returns from the cleaner.

Animals employ these odours a great deal during the breeding season to advertise their whereabouts, though it must be remembered that scent behaviour is instinctive, and not carried out with any purposeful intention.

It is a fact that the sight, or maybe the smell of a stoat, appears to induce a form of paralysis in a rabbit, so that it is unable to make its escape. A ferret has a similar smell, but in the open the rabbit does not appear to be in any way incapacitated if a ferret makes its appearance. But when one is introduced into the set most of the rabbits will bolt to the outer world. There will, however, quite frequently be one rabbit which appears unable to make its escape even when it is in a clear passage, and is not, as is generally believed, caught in a blind end of an underground tunnel. It is likely that occasionally a rabbit, underground, does succumb to the ferret odour, for when dug out even if not badly injured, the rabbit appears as though drugged and unable to make its escape, even when the ferret is again safely in its bag.

I have noticed that if one interrupts the outdoor drama between stoat and rabbit and the latter runs back to its hole, within a minute or two it is outside again, sometimes pursued by an angry buck. The reason is that it is contaminated with a very diluted stoat odour picked up from the air, and in the eyes of its friends it is no longer a rabbit but an enemy.

The old buck does not succumb to stoat odour but will frequently rush at a stoat, jump over it lashing out with both hind feet, often flinging the stoat a number of yards.

We say that a dog urinates on every post and against every wall that carries an attractive odour because he wishes to mark out his territory, or have some memento which will advertise his visit. In reality he wishes to do nothing of the sort.

He cocks his leg quite instinctively in response to a releaser, and he has no idea as to why he behaves in this fashion. The behaviour may fulfil a purpose, but it is one nature designed for him and it is not behaviour to which he consciously contributes. When the smell of female urine on a pavement plays the part of the releaser it provides a stimulus for urination, and as the anatomical formation of the dog permits him to pass only a very small quantity of urine each time he "cocks a leg," it becomes necessary for him to find from twenty to twenty-seven releaser smells during the course of a walk if he is to return home with an almost empty bladder.

It is this same releaser which operates when a visiting bitch (sometimes a dog) has passed a few drops of urine on the drawing-room carpet. Your dog - a model of successful house-training - is quite unable each time he encounters its odour to refrain from cocking his own leg against that of the table. It is an instinctive response, the smell of urine being a powerful releaser.

Bitches are equally capable of smelling but do not concern themselves with uriniferous stenches in the way that male dogs do. But when the bitch is approaching her mating period some of the odours released in her urine will inform each dog in the village concerning everything to be known about the present condition of her love life. As at this time the bitch finds her bladder exceptionally irritable she is constrained to pass several small pools of urine at various stages of her journey. When her time of mating comes due all the dogs will trace the odours back to her home with the unerring instinct of bloodhounds. This explains why on one particular day out of the year, the owner of the bitch finds the whole male canine population of the village sitting on her lawn, or on the pavement outside her gate.

Deer and antelopes possess an antorbital gland situated in the skin above the eye and they rub this onto twigs and branches until the whole area becomes permeated with their particular odour.

The dog makes sure that his own body carries his smell by urinating when possible among grass or leaves, then kicking his way through it using fore and hind feet. This anoints his limbs and all the underside of his body, particularly if he is low on the leg.

Some animals make use of faeces and urine as defence weapons. Apes in zoos become very efficient in the art and the male cat, as well as the lion and male tiger, can guarantee surprises if suddenly disturbed in their cages. Male cats are almost as active as male dogs in leaving their odours around their territorial rights.

The European bison, as well as the brown bear and the stallion, all use urine and droppings for territory demarcation, and the rhinoceros will build large dung piles, especially at the ends of his regular runs through the jungle. The bison makes a puddle in which he lies on his back so that he soaks his hump.

He then rubs this dry against trees on the line of his territorial boundary.

Monkeys, when angered or frightened, have the power of producing volumes of liquid faeces with which they splatter their adversaries.

Elephants can produce urine from their rear ends and volumes of water on occasion through their trunks, and they will direct both of these on unwelcome visitors. Camels and llamas will spit in one's eye, while bats, toads and frogs will all squirt urine in one's direction if disturbed or handled roughly. The skin of the toad also secretes a poisonous substance to which many dogs are highly susceptible. Vomiting, salivation and collapse may follow if a dog takes a toad into its mouth.

Many sea birds; gulls, cormorants, shags, and wading birds such as herons, will sometimes vomit if molested.

I was often amused when watching shags catching six-inch eels in the mud and sand of the Hayle Estuary in Cornwall. They would fill their interiors with live sand eels and if disturbed, would attempt to fly over the head of the intruder. When well-placed the bird would relax its anal sphincter and liberate a rush of liquid faeces. Incidentally, this would enable one or more wriggling, live eels to complete the passage through the shag's short intestine, and descend upon the visitor as effectively as though from a second barrel of the bird's armoury.

The large apes have a playful way of filling their mouths with water and squirting it at their victims (human, by choice) over quite long distances with almost certain aim. Quite a number exhibit what can only be supposed to be humour after a successful effort. They make loud, somewhat raucous, guffawing sounds, wrap their arms and hands over their heads - and bolt before retaliation descends upon them.

The habit of squids of squirting ink into the water as they go rapidly into reverse, induces the enemy to attack the cloud of sepia in the belief that it is the squid.

We humans can have but little appreciation of the highly important part of the scent glands, the odoriferous secretions, and the excretions of animals, play in their lives, since our own sense of smell is so poorly developed. When we realise how highly sensitive animals are to even highly diluted scents, it becomes quite apparent that in the animal world scent may be as important, sometimes more so, than vision.

In addition, the ears of animals are tuned in to recognise a great variety of sounds, both high and low in pitch, which we never hear. Dogs carried on a jet plane may go almost demented because they are compelled to listen to an extremely shrill supra-sonic whistle which on them has ear-splitting effect, although we do not hear it at all. They land in a completely exhausted condition and may be left with an intense phobia regarding planes, usually permanent.

Just as the animal's body may produce a variety of odours, so does that of the human being, though fortunately they are not easily recognised by people of our own colour. One of the comments made by black folk about white people is that they can put up with the individuals but can hardly tolerate their smell.

A similar comment is sometimes made by white people about the black folk, so it would seem that both smell, but it takes a person of the opposite colour to recognise the odour.

Animals are very conscious of the odours emanating from the human body and recognise very many more separate scents than we do, in much higher concentration. They are particularly aware of the smells derived from the breath, the ears, the axillae, the groins, the palms of the hands, the soles of the feet, and the sweat glands. The odours from all these parts are different and specific, and are readily identified and recorded by the dog as the smells which identify its master or mistress.

Curiously, dogs appear to disregard the artificial scents which ladies use to provide additional attraction, and can recognise body odours mixed with these as though the artificial scents did not exist.

Man produces a number of "sweat smells," some of which record the state of his emotions to the animal. One of the most easily recognised is the "fear smell" which emanates from man when faced by any animal which daunts him, or even by finding himself in any situation attended by danger. Most dogs allow themselves to be handled by experienced people, who produce no fear smell. The same dog which has behaved in a show perfectly with one handler, may be quite maladjusted in another, show, handled by a nervous person.

Children have special odours of their own, none of which appear to be offensive to animals. It is not until humans reach puberty that their odoriferous glands begin to secrete a great variety of scents.

Persons in constant charge of animals frequently learn to detect the approach, or presence of disease among animals in their care by noticing the smells which emanate from them. Cattle which give forth a smell suggestive of melons or pineapple are usually affected with ketosis, a metabolic disorder; while calves, monkeys and kittens, together with a great many young animals, may produce a very sour smell when their alimentary canals are out of order.

Learning in Young Animals

Children go to school and learn by listening to the human voice. To some extent young animals in domesticity also learn from being talked to in gentle tones, and undoubtedly learn the practical application associated with certain words, tones and sounds.

I have no doubt that adult dogs which have lived indoors with a family for a number of years also prick up their ears when certain words are spoken, words which they connect quite consciously with their own doings and affairs. Were it not so, there would be no occasion for me to take by own Scottie, Whisky, out into the kitchen and provide him with a plate of chopped meat and biscuits every time I wish to discuss plans for the next day of two, some of which would include whether or not to take Whisky along with us or whether we shall leave him in care of a friend. Also I find it unwise to refer to any member of the household by name in his presence, since Whisky immediately goes to his or her side and offers protection, while he cocks his ears and listens further in the hope of discovering if the persons mentioned is in my good, or my bad books.

This statement may at once cast me among the ranks of the anthropomorphists, an implication I shall strenously refute, for whether the scientists believe it or not, dogs which live as members of a family really do learn a great deal more than the animals with which the broadest-minded research workers are compelled to associate. Once again I will emphasise that to see into the minds of animals one has to live with them. One cannot really hope to learn a great deal from a cast-out racing greyhound, unvalued, unwanted, thrust into a cage, and given a number in place of a name, unless it be that one learns one should treat no animal in this manner.

In a state of nature young animals learn a great deal from play, and the instinct to play in animals is largely manifested during the period which terminates in complete adult development. Play exercises the muscles, and creates enthusiasm in such matters as acquiring speed, and learning the secrets of attack and defence. Even in childhood the sex hormones exert some influence, so that the little girl loves her dolls, and her brother specialises in hunting. He may commence with a bow and arrow, graduate to the catapult stage and through the air rifle merge into the sportsman possessed of some kind of shotgun. Children, by the sort of games they play, become conditioned to the life they will expect to lead in their adult years.

The herbivorous ungulates, ranging from the horse through the deer family to the giraffe, all practise the arts which it will be necessary to acquire if they are to survive when they leave their parents. In communities, the young of one female play first among themselves, and later with the young of other females. Even when they stray far from their homes, as lambs may do on a hillside, each one knows its own mother and her special call, while each mother can both visually discern her own young, and confirm this by smelling them immediately upon their return.

Sometimes play is devoted to special uses as when a swan teaches her cygnets to climb onto her back so that she can transport them downstream, or a shrewmouse teaches her young to hold each other by the tail in line while the leader hangs onto its mother's tail. In a similar way young elephants in the jungle

clearings will hold the tail of the elephant ahead of it in case one of their number should become lost.

Egyptian ichneumons also adopt a follow-your-leader system so that a family proceeding through scrub may sometimes be mistaken for a large snake since the undulations practiced by the father, who leads the way, are imitated by each young ichneumon in turn.

Not only do carnivorous animals such as lion cubs, kittens, puppies and young foxes play games suggestive of hunting, but they are also taught in the course of such gambolling how to catch and kill. It is highly important that they shall catch with this intent, instead of being eager to let their victim run indefinitely, so that they can repeat their play. As I have already mentioned, few carnivorous animals actually kill before they have received a considerable amount of maternal teaching.

Animals do not rely entirely on instinct, or on reflexes, but develop variations of what would otherwise pass as pure automatism, by a process of learning.

In 1894, Lloyd Morgan, then one of our leading animal psychologists, being apparently somewhat disgusted at the anthropomorphism displayed by some other psychologists of that period, mainly foreign I should add, formulated his law. Ever since, this has been known as *Morgan's Law*. To put it briefly, it stated that the actions of any animal must never be regarded as being the outcome of a higher psychical effort, but must always be ascribed to simpler, elementary causes.

Possibly Lloyd Morgan was a little too keen on ridding the study of animal behaviour of false sentimentalism, and as a result, animals exhibiting levels of intelligence exceeding the average (and there are a number of these) received no credit for their mental acuity, purely on a matter of principle, while a number of half-witted human beings, devoid of all sense of social responsibility, were at the same time being accepted universally as intelligent animals capable of reasoned behaviour.

One of the sensible things every wild animal learns to do, is keep as far away as possible from mankind.

But man is multiplying constantly and seeking out fresh territory, and as he advances the animals recede. In our own island of Britain there will soon be no room for animals, wild or domesticated. Horses are almost extinct with the exception of those kept for pleasure and sport - the working horse is a rarity. The towns become more and more overcrowded, people live in skyscrapers to enable more people to be maintained on every acre of ground, the pavements are thronged and to the dog on a lead the world can seem to be little more than a danger zone packed with thrusting legs and crushing feet.

Fortunately at the moment we have space for our cattle, but as towns constantly encroach upon the country it cannot be a great while before they, too, begin to diminish in number.

Waste grounds, where a dog can run free and relieve his physical needs, have jumped to ten times their original value and are snapped up as sites for factories and skyscrapers, and more multiple stores.

Very soon the cat will oust the dog as a household pet. It is a more accommodating animal and knows how to keep clear of traffic, refrain from soiling the sidewalks, and carry on its normal love-life when humans are asleep in their beds.

On those other continents where animals still live in their natural wild state, man is penetrating equally fast, and the animals are being pushed nearer and nearer to extinction. As soon as controls are removed and animals may be killed indiscriminately, most of Africa's magnificent collection of wild animals will gradually disappear, while those varieties which possess intrinsic value in relation to their horns and hides, or even a solitary horn in some of the rhinoceroses, will completely disappear from the face of the earth.

Later, the birds will go too, and from then onwards the insects will come into their own. It is quite conceivable that insects, unrestrained by their natural enemies, may prove more dangerous to the survival of humanity than all the animals of the jungle, even if their number were increased a hundredfold.

A single aphis, the ordinary greenfly of our gardens, produces eleven generations of young ones every year.

Charles Bonnet, a famous naturalist of Geneva years ago, began to count the produce of a single aphis, but at the eleventh generation, during the first year, he gave up. The number of aphides derived from the solitary specimen was at this time 441,461,010,000,000.

It is likely that as soon as the world loses its animals and its birds and its whole ecological balance is upset, insects will destroy humanity far faster than the atomic bomb is ever likely to do. If the insects alone do not succeed, the insecticides with which man will fight them may prove equally as lethal as the insects themselves.

TERRITORY

It is frequently imagined that animals living in a wild state wander about from place to place, and that the fox we see crossing the common today may quite possibly be fifty or a hundred miles away from the common in a few days or a few weeks time.

While it is true that animals may change their territory from sheer necessity, it is seldom that they do so in ordinary circumstances so long as the need of that particular type of territory exists. Very few animals are nomadic by nature and most of them claim their own circumscribed area of land, water, or air, of which they know every inch quite thoroughly. This they regard as home, and any animal forcibly removed from its territory may readily die of nostalgia. This frequently happens in a few species when animals are removed from their homes even to fairly natural conditions prevailing in a zoological garden.

Not only do animals possess a territory of their own which they may have striven hard to secure, but they probably also have their living and sleeping accommodation, their bath and their lavatory, in just the same way that we divide up our own houses into separate compartments.

Very few animals soil either their living or sleeping accommodation. A pig living in a modern sty, with separate toilet, will rank among the most hygienic animals alive.

The larger birds, the rooks, pigeons, and members of the hawk family, will endeavour to keep their young ones unsoiled by building openwork nests through which excrement can drop to the ground beneath.

Probably the animals least clean, which in a natural state soil their beds, are those nearest to man - the gorillas and other large apes.

Birds generally speaking, are very clean and tidy in their habits, and both parent birds every time they bring food carry away the excrement of the young birds.

Contact

Animals may be grouped in a good many ways in regard to their attitude towards men and other animals.

Firstly, we encounter those which fear man instinctively, and absolutely refuse to make friends with him, or become amenable to captivity. Examples are the weasels and some of the martens, although the *common* marten has been tamed.

On the other hand, a great many kinds of animals are willing to accept man's influence and domination, or are at least willing to live with him in order that they may enjoy the food served to them by human hands, even if they do not place a tremendously high value upon human society.

The common domesticated animals fall into this group but there are many others such as foxes, badgers, hedgehogs, rabbits, squirrels and grass snakes, in this country, which become very tame; in addition to a great many birds which can be petted and handled even though they retain their freedom. I have known jackdaws, seagulls, wild duck, robins, tits and even common sparrows, which insisted on coming indoors to be fed, yet were in no sense confined. The greatest drawback to encouraging a display of affection from wild birds is that they never cultivate drawing room manners.

An intermediate group, consisting of animals which are neither complete technophobes (rejecting man), nor complete technophiles (accepting man), includes a very great many varieties which are not usually tamed but will respond to domestication if they are taken in hand from the nest, or when very young. In most instances such animals need specially prepared houses and very often suitable water tanks, or pools, for those with aquatic tastes. Animals of this type are so varied it is only possible to mention a few. These include otters, seals; birds of prey including owls; dolphins, various kinds of fish such as carp; field mice, toads and many others. I knew one lady who kept a large collection of huge spiders which fed from her fingers, and we have all heard of the exhibitor of performing fleas who maintained a number of these in domesticity, and fed them on his own blood.

There are very few animals which can be captured and easily tamed after they have attained adult age and savoured the joys of freedom; nor does it seem to me to be kind to attempt to confine these within a dwelling house. When taken at an early age they never suffer the pangs of nostalgia and become contented with their lot.

Animals captured when fully grown, even if they are not positively dangerous, will sulk and fret, refuse their natural food and often die, not from self-inflicted starvation but from the shock of being transferred into an artificial environment.

Animals also differ from one another in the matter of the tactile sense. Some accept domesticity and enjoy being pulled about, stroked and cuddled. They like skin contacts. Others hate them, and although they will accept food and appear quite comfortable in captivity they strongly resent any contact with the human hand.

Horses like being stroked and brushed, and a sure way to gain the goodwill of a rough colt, if one can approach him, is to gently scratch his withers with the finger nails. If the colt happens to be lousy, it will express even greater approval. Young horses fresh in from grass usually harbour lice or forage mites, which may account for the pleasure they obtain from gentle scratching.

Even animals as different as the common toad like being held in a warm hand and gently stroked with a tip of a finger of the free hand. The frog will have none of it and hops away in great indignation. Toads, by the way, become very tame in captivity. Hedgehogs are also fond of having a finger run through their spines and will often respond by uncurling and accepting food. Since they harbour countless fleas of a variety peculiar to the hedgehog at the roots of the spines, this may explain their liking for contact. Nevertheless, I do not think these fleas cause a great deal of irritation, and it has been said on good authority that destroying the fleas may result in the death of the hedgehog. There is, however, a possibility that these animals are quite as susceptible to the injurious effects of insecticides as the fleas may be.

Other animals which like to be stroked in a tempo suited to their particular tastes, include pigs, monkeys, koalas, dogs, guinea pigs and white rats.

A great many animals, including many birds dislike contact

of any kind whether by stroking or by actual bodily impact. Swallows on a telegraph wire keep a space of two to three inches between each bird; seagulls perching on the ridge of a roof keep a space between birds of from twelve to eighteen inches. Waders, such as knots, congregated thickly on a sandbank, stand in formation and move about almost shoulder to shoulder, but never quite making positive contact with another bird. Storks consider it correct to spear with their beaks any other stork settling within range.

In contrast, farmyard hens, and chickens will crowd closely together for warmth, and sparrows and starlings in winter often sleep in a recess in layers one on top of the other. Budgerigars and parakeets are especially fond of close contact; so are monkeys. Pigs love being scratched and stroked and congregate very closely. Pigs are fattened nowadays at a temperature of 70° F. and the rule is to fill each house until there would be room for only one more pig. Conditions such as these appear to approach the porcine Nirvana – for as long as it lasts.

There is insufficient space in this volume to contain a full account of the types of territory favoured by the various species.

The area of a territory varies inversely to the concentration upon it of available food material. When foodstuff is plentiful the distance travelled by the owner is less than when food has to be sought far afield.

Many territories are communal such as those shared by badgers, or even rabbits, different families of which may share the same set, though using different parts of the interior.

Some badgers' sets have been in use from days long forgotten, and have been the birthplaces of countless generations of badgers. Although it is not generally recognised, a field may contain a great many families of moles in runs, which eventually all become continuous.

A friend of the writer owned a field which was riddled with mole hills. In an attempt to flood them out he inserted a hose-pipe in a mole heap on the highest ground at the very end of the field, and commenced to pump in water. Within a few minutes a great many moles came into the open, but at the other end of the field.

Badgers among themselves are good natured, sociable animals, and they can become on quite good terms with foxes, so that it is nothing unusual for a dog fox to share an earth with one or more badgers when it would not do so with another dog fox.

A dog's territory usually consists of the house, and the garden in which it plays, though the latter may be represented by only a path from the front door to the gate. Alternatively its territory may be a kennel and a square of concrete, or it may be the whole area of a large warehouse. Whichever it is, it still represents territory to be guarded, in which the dog is supreme when left in charge. The prestige of the smallest dog within its own territory far exceeds that of the largest which may be tempted to intrude, and the large dog when discovered and

attacked by the midget is in honour bound to tuck its tail between its legs and bolt as though pursued by a savage lion. The little dog will be particularly careful not to put a single foot over its territorial boundary, for outside that his mantle of prestige would slip off, and he would become a little dog again. The extent of the territory will have been determined from the moment when he first took it over. The boundaries will have been liberally sprinkled with urine applied at nose level to every available landmark, and by kicking the urine through any grassy patches, using both fore and hind feet in doing so, the little dog will have scented his body with his own particular brand of perfume, a fact which will stamp him in the nose of any dog doing a little prospecting in the locality as the owner of the similarly scented area of land.

It is interesting to note that dogs pay no attention to the solid excrement of another dog, even though they go crazy on discovering the faintest odour of his urine. Apparently faeces have no sexual meaning and undoubtedly the fascination of urine lies in its sexual implications since a dog which has been neutered before attaining puberty attaches little or no importance to uriniferous odours, although a dog neutered *after* attaining adult life continues to appreciate its attractions.

It is believed that an accumulation of faeces on the outskirts of any territory whether that of a dog, a bull, a stallion, or any of the ungulates, odd-toed or even, helps to warn off would-be intruders. Zoo animals, after their cages have been scrubbed, are unsettled until they have again anointed their boundaries.

As a general rule it is the males of the ungulates which patrol the outer boundaries, the females remaining always nearer the centre. The bull who is the dominant member of the herd prefers to stay in the centre with the ladies of the harem and drives the young bulls out to the boundaries, where they may deposit their droppings. The young bulls are by no means displeased since isolation sometimes puts them in contact with a straying female, well out of sight of the chief.

Territories are often divided up into areas set aside for excrement, feeding, fighting, or mating. Many ungulates, including red deer (but not roe deer), possess the *pawing instinct.* When they are not feeding they will congregate in a sheltered spot, often under trees, and paw the ground, throwing out the soil behind their bodies. As soon as one piece of ground has had the crust removed by this means, others will use it as a pawing place and will soon dig a pit large enough to lie down in. When rain comes the pit will develop into a mudbath in which the deer will wallow. This helps to destroy lice and as a long-term proposition it improves the coat.

Cattle and deer also endeavour to stand in water, at least up to their fetlocks, during the period in spring and summer when the gad flies are active. These flies lay their eggs in the animal's heels and standing in water acts as a deterrent.

The young animals, a few months old, show the same fear of the gad fly. As the flies cause no pain, but emit a buzzing note, it is strange that cattle of any age should dread them. They certainly do not know that in a few months time the eggs laid in their heels will have become converted into fat, inch-long, spiny larvae, boring a hole in the skin of their backs in order to escape to the open, to turn into another generation of gad flies. It is apparent that the behaviour is instinctive and that the releaser is the high-pitched whine produced by the wings of the female gad fly.

Deer and antelope also, anoint the trees bounding their territory, or within it, with the oily, odoriferous secretion of a skin gland which opens by a duct above the eye. It is known as the antorbital gland.

It would seem that in the animals' world although vision is essential, scent plays an even greater part in setting the pattern of animal behaviour.

Most of the pig family, even the domestic sow when left to her own devices, will make a bed beneath a tree trunk, or in a hollowed-out bank, or may make use of a hole dug by some other animal, which may be enlarged if necessary.

The behaviour of a farmyard sow, "gone native," is rather interesting. When she has piglets underground and has been out foraging, she will run towards the hole on her return and dive in head first to assure herself that her piglets are safe.

After this she will come out again, sniff the air, then turn round and re-enter the hole in reverse position. This is a display of the natural porcine instinct to enter a hole backwards, ready to attack any possible intruder inclined to follow her into the hole.

Most animals have a series of runs, or pathways, through thick cover, or through the jungle, through which they can make a rapid and unobserved approach or retreat. In similar fashion tracks run across plains, or fields, to other places which it may be necessary to make for in a hurry, and most of these are in regular use.

Animals caught napping away from home, recognise these runs as the roads to safety, more especially as they commonly take advantage of any cover available en route.

Quite frequently, in the rabbits and rodents for example, the runs may lead through one or more underground burrows and out again at their other end, while the regular runs of foxes and badgers often pass close to piles of rocks which might provide shelter in emergency.

Birds are very particular about their territory especially during the nesting or pre-nesting period. Both cock and hen may play their parts in its guardianship, the cock driving away other cockbirds, and the hen attacking other hens who might have designs on her mate, as well as the territory.

Every cockbird, who is in the mating mood, seeks out the territory he proposes to adopt while he is in a definite physiological condition, one which induces in him a programme of behaviour he would exhibit at no other season. The reason for all this transformation in his personal appearance, as well as in his behaviour pattern, lies in the changes which have recently taken place in his gonads, the male sex glands. Lying hidden in his abdomen, these are enlarging and throwing out their hormones into his blood, and after being virtually a neuter throughout the winter he is now blossoming out again into full masculinity. He may be driven by his increasing excitement to set out on his marital journey a little early. Then, in company with a few other cockbirds for whom as yet he has developed no antipathy or sense of aggression, he will fly about in bachelor state on quite friendly terms. But not for long. Irretrievable changes are taking place in his temperament. Each of the male birds will begin to eye the others askance, since they are simultaneously becoming smitten with the same idea — the acquisition of personal territory, a nesting site.

One or more of the birds may remember, or instinctively make for a site, which has served his purpose on a previous occasion, and he will enter this and fight off all others who attempt to approach it.

The cockbirds will attempt to perfect their songs, not consciously perhaps, for unlike the operatic singer they cannot, as it were, step back and consider how good or how ill the attempt sounded, nor can they record their efforts as though on a tape machine, and endeavour to find out how their songs sound to other ears. But the more they sing the more antagonistic they become, for the object of the song is not to bring pleasure to human ears but to carry out two important objects; the first to drive away other males by suggesting that the bird is as indomitable as his song is loud and clear, and the second is to inform a possible mate as to where so virile a husband is to be found, and to persuade her to single him out from all competitors.

He will now fly at his former companions with beak wide open and will assume the posturing of aggression characteristic of his variety. He will develop a special hissing note, which coming through his beak may suggest to his rivals that he is half-snake as well as bird. His breathing will increase in rapidity, and as he visibly pants his feathers will stand on end as his skin contracts, until he resembles a puffed-out ball of feathers far greater than his usual size — a terrifying spectacle to another bird who has hardly reached this same stage on the road to full virility.

With all this display it will not be long before a hen bird will put in a modest appearance. At first the cockbird will hardly be certain whether the newcomer is male or female, for although birds have an eye for color (just as dogs *could* discern sex characteristics visually if they had better brains), cockbirds still seem unable to decide that the bird in gaudy trappings is probably male, and that the drab little bird which has just arrived is probably female.

To make quite sure our bird will posture and put on his show of aggression, watching, of course, to see what the reaction of the new arrival may be. If it is a similar display of aggression, mingled with mild apology, the visitor is a male, to be shooed away. If, however, the response is shy and maidenly, though mingled with an air of invitation, the mate has arrived.

Bird territory is the area which a pair of birds can defend against all other avian intruders, and it may be regarded as their own food-preserve, combined with an approved nesting site.

After the mating, the fury abates a little, and the aggressiveness of the cockbirds one to the other tones down somewhat. When the egg-laying has been completed it is by no means unusual for another pair of birds to make a successful attempt to settle in close proximity. This will depend to some extent upon the food requirements of the two pairs. They may not be identical, or it may be that both pairs are caterpillar-minded and that in the fields, or immediate vicinity, caterpillars abound in great quantity. This applies especially to market gardening land and farming districts.

In cases where the food requirements are only sufficient to provide for the one pair of birds and their family the resident pair will go into battle so far as their duties will permit, and drive away the intruders, and in this they may be assisted by other birds nesting in the locality which may show co-operative instincts in such circumstances.

Birds which do not sing may yet produce a great deal of noise, which answers the same purpose. In a great many cases these songless birds are those which more often remain more or less mated in a companionable way throughout the year, while song birds are often more changeable, especially as so many of them migrate. The life of the song bird is not usually a very lengthy one and migratory birds run small chance of retaining the same mate for long, while resident birds may do so more easily. However, there is no fixed rule in affairs of this kind. Song, nevertheless, is not an aesthetic commodity, it is introduced for a purpose, and if we accept the fact that the mute swan, for example, does not sing we may presume it is because it will not experience any need to do so in order to attract a mate.

The territories of different kinds of animals, including birds of course, may overlap, so that foxes, badgers and blackbirds may share the same area. Although the food of the fox will differ considerably from that of the blackbird, that of the bird and the badger may be more alike. But none of these animals, not even the badger and the fox who have food habits even more alike, will quarrel, since the instincts of any one of them do not include the others in the list of animals to be attacked and driven off.

Two male badgers, two dog foxes, or two cock blackbirds in the same territory would be quite a different story.

It is, however, not unknown for a thrush and a falcon to share the same territory quite harmoniously, and while the falcon partners will cheerfully bring home thrushes, young and old, to pull to pieces for their youngsters, they would not dream of attacking the neighbouring thrushes which are protected by territorial rights and amenities. But a time may come, when one of the young thrushes on his trial flights may overfly the boundary. I will refrain from harrowing the feelings of my reader with further details.

One must distinguish between mating territory, nesting territory and certain adjoining areas, which may be grouped as feeding territory.

The woodpigeon, for instance, may have its own group of trees as mating territory, its special tree as nesting territory, but it may find it necessary to fly a short distance to what may well be communal feeding territory, which it may share not only with other pigeons but with rooks, starlings, small birds of many kinds, possibly with some pheasants and partridges, a number or rabbits and a brace of hares.

This sort of thing will continue when the reproductive organs of both sexes have reverted to their neuter character, and a good many of the animals mentioned above will continue to associate in pairs, or to remain on friendly terms until such time as the breeding cycle again comes into play when they may repeat the whole pattern of behaviour.

Some of the seabirds, for example, may include a vast expanse of sea in what may be considered feeding territory, just as a herd of deer, or buffalo, or elephants, may roam a circumscribed area of forest. These larger mammals are not bound very closely by any restrictions but the seabirds will have to carry out an organised timetable to keep pace with tide and weather conditions, and their lives may necessitate close successions of flights over long distances between land and sea.

We speak of *local orientation* when we refer to the robins which have their own nesting territory and leave this later for a fresh one, although they may cover no great distance, as miles go, in their migration. Birds which migrate to other lands exhibit *distance migration.*

The more one studies animals the more evident it becomes that their life is spent in two compartments, the one in which nearly everything they do is instinctive and carried out in response to a stimulus or releaser, tempered with a variable amount of conscious adaptation of behaviour in keeping with life's circumstances. In the other compartment their life is even more pre-ordained, since under the influence of sex hormones their whole behaviour is almost mechanical in its performance without any sign of variation, every stage of the journey being arranged in advance and each stage leading by regular sequences to the same inescapable termination.

THE SOCIAL HIERARCHY

Whatever idealists may regard as a right and proper attitude between human individuals, it is quite certain that in the natural state of living true socialism is unknown, and class distinction among all varieties of animals is fundamental. It has always existed and will ever exist. Among animals in more exalted spheres a true regard for differences in the social grade of individuals minimises fighting. If all animals were rendered equal in their social position they would quarrel until they had again created a hierarchy in which they descended the rungs of the social ladder from the despot to the slave.

The sorting out of social caste among animals is always attended by bloodshed and slaughter, but it is an essential operation whenever animals congregate. The animals at the top of the ladder are biologically superior to those lower and lower down the rungs, very often in physique as well as in general well-being.

Here we may find a state of affairs different to that existing in human society.

Where intelligence, cunning or experience lift one man up into a high position he may in a physical sense be puny and far inferior to those beneath him in physique, although his mental processes are more active. This applies especially to an organised society which depends on its brains rather than its muscles. In the working-class world the bigger and stronger the navvy* the greater is his earning capacity, provided his bodily physique is also accompanied by the will to put his strength into active operation. Usually the navvy is a member of a union which is organised to serve his interests and secure for him adequate wages and the best possible working conditions. The head man, controlling the activities of the union may be so frail that any one of the navvies could dispose of him in unarmed combat with one hand tied behind his back. Nevertheless, this delicate individual may exercise a ruthless despotism, and his authority and position will be acknowledged by every navvy in the union. None the less, a member of an aristocratic group may possess the physique of a navvy, but he does not use his muscles to earn his bread. Incidentally, his mental capacity may be inversely proportional to his muscular strength.

The difference between man and the lower animals is exemplified here since man places brains before brawn, while in the case of animals, might is always right. Among animals a natural tendency to leadership goes always with superior physi-

* The term "navvy" implies no disrespect but is used here to indicate any man who works entirely by the use of his muscles.

cal strength. What underlies leadership is uncertain, but it is probably associated with hormonal output. We can make hen canaries sing and dominate the cockbirds by injecting them with male hormones, and it is likely that a combination of hormones as supplied by nature might do even better in determining leadership.

In human circles leadership is usually accompanied by greater mental ability or greater wickedness.

The social status of animals is most easily studied in the poultry yard. Similarly, one may watch behaviour within a herd of dairy cows lined up at the field gate waiting to come in to be milked, as well as when they return to the field. There is among cows, as in hens, a definite order of going and coming in accordance with social rank, and woe betide the cow who endeavours to usurp a position higher than her own.

David Katz in *Animals and Men* (a Pelican book), cited the following example in the pecking order for a group of seven hens:

A	pecks	BCDEFG
B	"	CEFG
C	"	DEFG
D	"	BEFG
E	"	FG
F	"	G
G	"	none.

Note the slight inconsistency where D pecks B, instead of B pecking D. Here something must operate other than sheer physical superiority.

Note that the cockbird is pre-eminent, and that his rank is conditioned biologically.

Schelderup – Ebbe discussed a triangular order where:

B	pecks	C
C	"	D
D	"	B

When hens are first introduced to one another the more assured and fearless hen (undoubtedly stimulated by her hormonal output) at once pecks the other hen. This relationship persists indefinitely unless some accident or illness befalls the fearless hen, when the position of the two may be at last temporarily reversed.

When two equally fearless hens meet and neither will submit to being pecked by the other, they go through a ceremony of wing-dropping and commence to fight. This may continue for only a few minutes or it may go on for a quarter of an hour by which time both birds will be bleeding from wattles and comb and suffering from exhaustion. It is very rarely that one bird will peck the eyes of the other. Eventually one hen will walk away a few steps, hoping probably the other will not follow, and this usually ends the matter for the moment, though it is likely that battle will again ensue before the matter of rank is finally settled.

When a strange hen is introduced into the flock she is at once pecked by all and sundry and probably chased if she will consent to run. But, if the cockbird is attracted by the stranger and treads her, the new bird will immediately develop priority status, and this will persist until the cockbird transfers his attentions to other hens. The newcomer then reverts to her original much-pecked position in the flock.

In some instances the newly introduced bird may be one which held high rank in the flock from which she came. Instead of submitting she, too, may stage a battle royal, out of which she may emerge with flying colours. Such a hen exhibits a high ranking potential, combined with spirit and intelligence. Probably she may never become the matriarchal despot, but she may earn the privilege of pecking a number of her would-be peckers.

It could easily happen, too, that such a forceful character might find the combined activity of the other members too determined, so that she might easily end up by being killed.

The social hierarchy I have described in the poultry yard exists similarly among most species of animals which congregate in flocks or herds.

One finds among both *individual dominance* and *species dominance.* Geese for example may torment ducks and hens, while a goat considers itself equal in rank to the cows it leads, and superior to the sheep.

The farmyard dog dominates the cats as well as the cows, but the bull may not be so easily subdued.

Peewits attack crows on sight, and the latter seldom offer a great deal of resistance. This antagonism is associated with the liking which crows have for plover eggs.

But males do not always dominate females and in the non-breeding season the ruling within the majority of flocks and herds, as in the deer family, is matriarchal. In domesticated cattle which, like the domesticated human female enjoy a breeding season lasting throughout the whole year, the ordinary rules of procedure fall to the ground. The farmyard bull retains his dominance, which in his case is not a seasonal prerogative, while in the human race the male usually gives throughout the year the orders with which he has been inspired by the female.

There are countries, however, in which the males show less shyness, and submit quite openly to female domination. Perhaps in the long run it makes very little difference to the results achieved. The reason behind matriarchal rule is simply that the female produces more of her female hormone than the male does of his own particular variety. As I have mentioned elsewhere, both the male and female hormones are mixtures of each in varying proportions. It is quite likely that the dominant female, in addition to her feminine charms has, also, just that faintest trace of masculinity thrown in.

It is only when his gonads are particularly active and his output of hormone at its maximum that the male asserts his superiority and the female becomes sweetly submissive.

In the animal kingdom the female is more capable of taking care of herself than the male could ever be. The latter has only one really useful purpose, which he can put into operation through only a small part of the year.

In any case there are always too many males, and if a few got gobbled down by hungry predators it is of no great concern to the rest of the herd. It is for this reason that the indispensable female assumes the central position while concentric circles of males afford her protection.

The carcases of a few males on the outer fringes satisfy the wants of the marauding animals, a worthy sacrifice if the female is spared to produce more and more females, in order that the species may survive.

The Pack Boss

The domestic dog stands out from all other mammals as the animal most capable of making itself completely at home in the family circle — of which it regards itself as a member.

No dog ever sees a great deal of its own body, nor does it appear to recognise it easily when it gazes in a mirror. There is nothing, therefore, to lead one to suppose that any dog pictures itself as the prototype of any other dog it sees, especially as every dog looks different to all the others. It is therefore likely that most dogs reared in a house from puppyhood become 'imprinted' and if a dog experiences any curiosity at all about its own appearance, which is unlikely, it imagines itself to be just another human being like all the rest of the household. An extrovert is seldom greatly bothered by personal appearance.

Normally, every house dog recognises one person as leader of the pack, the "Pack Boss," and this is usually the dog's owner. The dog itself in company with the other human members of the pack obey the orders of the "boss," while in addition there is a set of unwritten rules which all must follow.

The most important of these insists upon absolute and complete loyalty to the pack boss, and the pack.

It is the dog's especial duty to act as a policeman. He must guard the house, its contents, and all the members of the pack by night and by day. In return he gets full board and lodging and the affection of the pack members. If the board runs short he will thrive on affection.

There still remains in every housedog one marked example of his original pack instinct, dating back from the time when he was a wolf. When the pack was at home and asleep it was customary for certain individuals to remain awake patrolling the outskirts on the lookout for intruders. On the first suspicion the sentinel would sit down, throw back its head and howl dismally, a signal to all the others to join in, with the result that the intruder became aware that the pack was present in force, and promptly beat a retreat.

Nowadays, when one puts a disc on the radiogram and produces a sound reminiscent of a wolf howling, your own dog regards it as the signal to join in. The wailing is the releaser, and the primitive instinct to howl in unison cannot be disregarded

The Psychology of Feeding

It is generally recognised that animals feed when they are hungry and that men, hungry or not, feed by the clock. This applies to men and women who work set hours and have times allotted to them for meals. It would never do in industry if men and women started to eat and drink at any odd time.

Animals do not always eat when they are particularly hungry but when food it available, which may be at frequent or infrequent intervals.

The ruminating animals fill their paunches rapidly whenever opportunity arises and chew the food over again during their leisure.

The carnivorous animal makes a kill when hungry, opens the carcase and eats the liver of a large animal, or may swallow the whole carcase of a small one. If his kill is more than he and his mate and family can finish at one meal he may return to it on the following day, or even sometimes on the day after that, for carnivora, as a whole, are not unduly worried if their meat becomes "high."

However, by now the scavenger animals, which may include the ants, will probably have picked the bones clean.

Fish feed according to the temperature of the water and the amount of light reaching their optimum depth. They vary a great deal in different varieties as regards feeding habits. Fish which feed largely on the surface may experience periods of gluttony when "the fly is up."

Experiments with hens have shown that they eat more at one time when the amount offered is large than when it is small. If when the bird has eaten to satiation point the food is removed and a fresh lot offered, the hen will commence to eat again but will not eat so much as on the first occasion. If, when it stops eating a fresh supply is brought to it, it will eat again. In this way a hen can be persuaded to take some food from each fresh pile on as many as six or seven separate occasions. The total amount of food partaken of will weigh about one-and-a-half-times the original meal consumed. There is a limit to the amount of food a hen can stow away in her crop, which like the rumen of the cow is simply a storehouse which enables the bird to eat quickly while it can, then get away to a safe spot to digest the food at leisure.

When animals feed in company they eat for a longer period and swallow much more food than when they eat singly. The spirit of competition comes into force, and the pecking habit or its equivalent is manifested always when animals experience greed.

The male animal almost invariably places food first in order and sex second. The female, on the other hand, puts lovemaking in front of everything. The female in oestrus often fasts for one or two days until the mating urge has been consummated.

Animals have their likes and dislikes in the matter of foods and flavours, while birds and monkeys also exhibit colour preferences. Fowls dislike blue while many fishes like yellow but dislike red. Monkeys will select foods wrapped in one particular colour with great regularity when coloured wrappers are used, even when the contents are identical in each case. Nearly all my own monkeys chose red wrappings in preference to blue, and gold in preference to silver. They preferred silver to red in the majority of experiments. One of the attractions of silver paper appeared to be that by holding it up to the face the monkey could see a reflection of its own eye. A tiny piece of looking glass was examined in the same way.

Dogs, being colour blind, recognise only degrees of brightness but appear to disregard colour completely. When sweets wrapped in different colours are offered they grab the nearest. Experiments with dogs using food in coloured receptacles usually fail since the olfactory perceptivity of the dog is so marked it can smell food in what appears to be a closed compartment, and if any receptacle has been touched by a human hand, the dog can pick it out immediately.

Animals soon tire of any one variety of food if it is offered continuously and exclusively. My own dogs come to my room every morning and eat with great relish four cubes of fruit cake thrown to each one, for two to four days. After this they accept one or two pieces and refuse the third and fourth. Two days later they refuse the cake altogether but accept sweet biscuit with avidity. After several days of this they will accept no more biscuit but will eat cake again with great relish.

One of my Rhesus monkeys when offered a dish of assorted fruit would eat from it in the following order:

(i) Dried fig, (ii) grapes, (iii) banana, (iv) nuts, (v) apple, (vi) lettuce, (vii) biscuit.

The order seldom varied excepting that (i) and (ii) might interchange, which may have depended upon the state of thirst. When milk was offered in a cup (from which the Rhesus drinks in human fashion while most other monkeys lap like a cat) the milk was usually consumed at one effort before the monkey started on the fruit.

All this depends upon individual taste. Another monkey might choose a different order of selection, but the regularity of choice in this one monkey denoted definite taste preferences.

It is often stated that animals know what it is good for them to eat to overcome any existing mineral or trace element deficiences in their own metabolism, or any lack of a particular vitamin. It is also supposed that sick animals choose herbs which will help to make them better when they are sick.

It is a fact that dogs with digestive disorders will eat couch grass or any coarse herbage, and vomit later. Also dogs living on farms, which catch rabbits and become invaded with tapeworms as the result, will clear themselves of the parasites by swallowing mouthfuls or rabbit fur, which mechanically detaches the head of the worm from the intestinal wall.

Carnivorous females eat their afterbirths without apparent ill-effect and most even-toed ungulates ranging from cattle to deer and goats, endeavour to do the same. Sometimes these herbivorous animals get the afterbirth lodged in the throad or oesophagus, sometimes with fatal result. It is believed that eating at least some portion of the afterbirth is a necessary preliminary to the essential licking of the young which aids its respiration and establishes a link between mother and young. This applies to carnivora as well as herbivora and many apes.

Cattle which are unthrifty and lacking appetite often leave the fresh grass and make for patches of weeds or herbs in the hedges, or in strips specially planted with them. Cows deficient in phosphorus will eagerly chew dried bones.

But a good deal of experimental work has made it clear that animals have little or no selective instinct which tells them the herbs which contain the elements they lack when ailing or unthrifty, and even if they had the instinct it is unlikely they would find enough chemical in weeds to correct any deficiencies present.

It is nevertheless true that hens, which lack time to make shell, will eat the shells of other hen's eggs, or will pick out crushed eggshell when it is added to a mixture of grains.

Most animals deficient in lime will take it voluntarily in some form or other if it is made available. For this reason the farmer supplies minerals in the form of bricks which the cows lick.

At this point I would like to digress to show how feeding in animals, their likes and dislikes, may be linked up with a pattern of behaviour, with the result that if the pattern is altered the effect upon appetite may prove disastrous.

A rather remarkable example has just come to my notice and in recording it I would draw your attention to words I wrote on page 100 to the effect that one needs to keep an animal and study it at first-hand, and that a great deal can be learnt from the simplest pets.

You will recollect that I was surprised to find that my canary appeared able to discriminate in the matter of two packets of apparently similar mixed canary seed, eating one and refusing the other.

The reason has now become clear and I was probably

wrong in ascribing it to some difference in flavour of the two samples of seed. I have now discovered that the seed sold in green coloured packets and that sold in bulk, both came from the same consignment and were identical.

Before I bought the first batch of seed in bulk and poured it into the china food receptacle attached to the inside of the cage by means of a spoon, (the receptacle being reached by raising a glass window), I had in my own lazy way inserted the corner of the green cardboard packet through the opened window of the cage and with a lot of shuffling and rattling (which appeared to cause the canary some pleasurable excitement), I had tipped seed straight from the packet into the china recepta-

cle. When I fed from the packet all went well — but the canary refused to eat when the grain was taken from the bag containing the bulk sample and tipped into the receptacle with a spoon.

It is now apparent that the mind of the canary had over a period of time developed and accepted a feeding behaviour pattern. This was associated with a green packet making its appearance in the cage, accompanied by excitement and rattle. When I used a spoon the pattern changed and the canary did not approve of it.

I have since proved this by filling the green cardboard packet with the bulk seed which the canary previously refused. Offered now from the green packet it has become highly acceptable.

The rattle and the sight of the green carton were the releasers which were lacking when I used a spoon.

Simple as all this may appear, it is something I should not have observed had I not owned a canary, or if I experimented with birds fed by laboratory attendants.

The thing I would like to know, although I hesitate to try it, is how the canary would behave if it saw me filling the green carton from the bulk packet?

I do not think the mind of the bird would recognise the deception, but at the moment I shall not take the risk.

THE MATING INSTINCT

Women, sows and cows appear to be the only animals capable of mating the whole year round. In all other animals mating is a seasonal matter which happens usually only once a year with a respectable interval of anoestrus, or barrenness.

Domestication is the cause of this increase in the fertility rate of certain females, and it is probable that if we lived in a world of eternal sunshine, with a little rain three nights a week, every female animal would breed the whole year round and we should all die within a couple of years from sheer overcrowding.

Mating in the wild is so timed that the young ones will appear in the season of the year when it will be warm, with plenty of food for the mother, and for the progeny when they require it.

Mankind have the advantage of feeding well during both

winter and summer and of being able to sleep warmly throughout the whole year.

Woman completes an oestrous cycle during each month and continues to do so throughout the year. Each cycle and part of it is manifested by a pattern of behaviour and temperamental changes characteristic of the phase. A vixen behaves in precisely the same manner but she does so only once in the year while the human female repeats it thirteen times. One can see how right Sir Walter Scott must have been when he described woman as "variable as the shade."

Domesticity implies an equable temperature and regular food throughout the year. Most wild animals which have accepted fellowship with mankind appear to increase the number of oestrous periods per year to conform with the new conditions. It is a slow change for many bitches still only have one oestrous cycle each year, though the majority have two. Usually the smaller dogs, which are petted and kept almost wholly indoors, have more "seasons" than bitches, let us say of collie type, which work in all weathers in the open air.

Each oestrous cycle may be divided into:

(a) Pre-ovulation, accompanied by a desire to flirt and be courted.

(b) Ovulation, attended by a desire to mate.

(c) A luteal phase with quiescent ovaries. A coolness is exhibited towards the opposite sex.

(d) Complete anoestrus wtih frigidity.

Then comes a return to (a). When fertilisation occurs the luteal phase (c) persists for a considerable period and in the females of a few species it lasts until parturition takes place.

A great many females in the wild state show oestrus in spring and early summer but others, such as the vixen and the hind, develop oestrus late in the year. The hinds of the red deer start their "rut" at the end of September and by the first week in October the polygamous stags are in the midst of their crazy season, marshalling their harem, chasing off interloping stags, and generally after a fortnight merging into a state of unkempt exhaustion attended by semi-starvation due to inability to eat during this period of excitement. In a few instances stags may keep up the pantomine into early December.

Cattle in the wild state usually show oestrus late in spring and if they do not at once become pregnant may repeat the cycle occasionally, until June or July, so that following a nine-months gestation they calve between February and May.

In domesticity cows show oestrus every three weeks during the spring and summer, and usually throughout the year, although their fertility ratio is higher in the summer than in the winter months.

The human male does not undergo a regular cycle as the female does although the poet has told us what happens to a young man's fancy as Spring approaches.

The wild animals certainly do approach the neuter condition while the females are in the barren or pregnant condition, and it would seem that the activity of the gonads is stimulated by food and by sunshine. The male is not attracted by the anoestrous female and it is not until her pre-ovulation stage that she produces the odours which stimulate the male to return to full masculinity. Some of these odours can attract a male from a considerable distance as is evidenced by the female moth, and although mammals have not such keen olfactory perceptivity as moths, they may yet be attracted from a considerable distance.

Courtship is as well-established a custom among animals as among mankind, although in animals it is conducted with greater dignity. It was instituted for the benefit of the species rather than for the delectation of the participants. In every case it is the female who sets the pace and the male who blindly keeps step with her inclinations.

When the female odours have induced his gonads to arrive at their optimum activity, the male carries out the type of posturing and the general behaviour peculiar to his particular species. All this is designed to encourage the female to reach the stage of ovulation.

Although the males, particularly among the birds, put on bright colours, it must be remembered that although it is possible that such flashiness may aid subconsciously in building up the fertility of the female, and in hastening her egg production, it must not be presumed that she admires the colour scheme of the male plumage, nor that she is even conscious that it exists. It is quite certain that cockbirds do not recognise hens by their dowdy garb, and I have already described the appression display put on many male birds even when it is a hen that puts in an appearance. It is likely the bird "sees" a difference in the plumage but its brain does not lead him to understand the association between plumage and sex, in much the same way that a not-impecunious husband "sees" that his wife is wearing a new hat, or a new frock, but the fact does not, as he would put it, "ring a bell," until his wife verbally draws his attention to the fact, or he receives an account from the milliner or the dressmaker. It would be very unwise to attribute to any male animal a greater degree of intelligence than is possessed by the human male.

There is one marked similarity in courtship the whole world through, whether the participants be a brace of spiders or a couple far higher up in the social scale.

It lies in the fact that every potential wife must be fed, and brought into good, sound breeding condition, and this is considerably assisted by presents of food made by the male partner to the female.

It is instinctive in every male, greedy and selfish as he may be in his ordinary daily life, to behave in this generous manner when in mating mood, and it matters not a jot whether he be bringing a fly, previously sucked dry, then wrapped in layer upon layer of spider web, or whether the gift takes the form of a box of chocolates, or champagne and oysters, the principle involved is precisely the same. When the female is agreeably disposed she will accept such presents, for their token value if for no other reason.

Just as we males progress from diamond rings to house-hunting expeditions, so do the other animals proceed from presents of food to the acquisition of personal territory with nursery accommodation.

Courtship, when it begins in a flock or a herd, proceeds apace and takes on all the appearance of an epidemic. Every female is intensely interested in affairs of the heart. In a herd of heifers the first sign shown by any one of them that she has a date with the bull, throws all the other heifers into a flutter of excitement. They crowd around, lick her, try to mount her and no doubt by one means or other swallow some of her exuding hormones. It will not be long before every female of the party develops similar symptoms.

In human analogy one may compare the excitement arising among a roomful of girl typists when one of their number arrives one morning shyly brandishing a new engagement ring. One may regard this as a suitable moment to advertise for a fresh batch of typists since the epidemic spreads among typists nearly as rapidly as among heifers, and it will not be long before there will be additions to the list of vacancies.

Incidentally, I might mention that in the days before hormones were fully understood, long before they were marketed for purpose of injection, farmers were in the habit of dosing their young sows and heifers with the urine of a pregnant cow, or one showing oestrus, in order to bring them rapidly into breeding condition.

There is insufficient space in a single volume to discuss all the patterns of breeding behaviour in animals. I have made considerable mention of this already in *The Female of the Species* (Country Life, Ltd.), while those particularly interested in the mating patterns of birds might well read Dr. Stuart Smith's *How to Study Birds.*

MARRIAGE

In wild life animals appear to select a mate and live together quite happily. Fortunately they talk very little, are not obsessed by too lively imaginations, nor are they unduly disturbed by a little promiscuity so long as it does not interfere with the general pattern of marriage and the bringing up of family.

Even the little female sparrow, living happily with her mate, may be waylaid on a foraging expedition by a bevy of bachelor sparrows, and nobody considers it necessary to rush off to her mate with a garbled version of what went on in the apple tree. In the same way the hen pigeon may indulge in flirtation with a couple of lonely males on the ridge of the roof while the cock-bird sits dutifully on the eggs in the rafters beneath. If the hen returns straightway to the nest and carries on with the incubation she will not spend her time worrying whether some feather-brained female is trying to seduce her husband while he is out scouting around for a breadcrust.

In the animal world what the eye misses the heart never grieves over – and even when the female eye does observe, the hen will never lay the blame at the husband's door but will relieve her feelings quite satisfactorily by flying at the seducer and pecking her eyes.

Different kinds of animals attach different values to the same things. Nor are animals concerned so much as men and women are about whether a prospective partner is likely to prove suitable. While a girl seeks someone tall, dark and handsome, the female elephant is not bother at all so long as the next elephant that comes along turns out to be a bull. If he is a hundred years old or a mere child of twenty doesn't matter either.

There are no old ladies in the animal world. Animals suffer no change of life and breed until they die of old age.

Male animals are interested only in odour and its implications, they probably take little interest in the female face and recognise a mate not by her features but by the way she smells.

A virile stag would as soon mate with an established great-grandmother as with a gentle maiden hind – providing she were equally attractively scented.

Animals vary a great deal in the duration of marriage to the same individual. Foxes, for example, stick to one mate year after year, and it is the general complaint of breeders of silver foxes that they have difficulty in persuading a vixen to form a new partnership when it becomes necessary to do so in order to breed better skins.

The dog and the male cat are the world's greatest philanderers for after a brief honeymoon the male departs never to return, leaving the female partner to keep herself and her family to the best of her ability.

Wolves as well as foxes all remain happy with the same mate, probably throughout their lifetime. They look after their mates and bring in food for them and their families.

Badgers are also faithful animals but as they do not usually carry food underground and are mainly vegetarians with the addition of a diet of worms, beetles, insects, slugs and snails, they take turns in looking after the family while the other forages for food for itself. Both the male and female badger are devoted parents. The male wolf, however, is rather partial to wolf cub, but not when the mother is in sight, which she contrives to be whenever the father is at home.

It is, however, always difficult to decide whether the pair of animals one sees in any particular place during one year is the identical pair one saw the year before. A great deal of bird-ringing and marking animals has been carried out. It would appear that many of the insectivorous birds as well as members of the swallow families, and also the predatory birds, remain attached to the same partner, although as the expectation of life in birds is not prolonged it follows that the acceptance of a fresh mate is often inevitable.

Many of the rodents, rats and mice for example, incline to constancy but here again life is very uncertain.

David Katz in *Animals and Men* (a Pelican Book) includes swans, doves, grey geese, wild duck and the rhinoceros, in his list of animals which pair and remain constant throughout life, while other writers mention cougars, robins, wrens and hedge sparrows.

The *rearing of the young* is usually the mother's task though often the father helps, and in rare cases may accept full responsibility while the mother enjoys herself or seeks another mate for her next marriage.

The male sea-horse possesses an abdominal pouch which reminds one of that of the kangaroo in miniature. He swims about, inviting females to lay their eggs in his pouch. When he has collected a couple of hundred eggs he fertilises them. Forty five days later they begin to hatch out. The interesting fact is that the male temporarily becomes feminine. His abdominal pouch takes on the functions of a communal uterus and in each of its numerous honeycomb compartments an egg lies amid new tissues developed from the pouch lining. Each of these takes on the form of a placenta and when the yolk of the egg has been absorbed each embryo sea-horse is ready to face the world. Sad to relate the male always seems to suffer a difficult parturition. He buries himeslf in the sand and displays every outward sign of suffering. Eventually he shoots out batches of young sea-horses with explosive force but it may take two or three days before the whole operation is completed.

The stickleback makes love to his numerous mates while standing on his head, his lips meanwhile clearing a way in the sand to receive their eggs. When one female has deposited her eggs, the male seeks another and yet another willing fmale until he has as many eggs collected together as he deems sufficient.

The releaser for this instinctive response is the sight of the corpulent belly of the egg-bearing female. Experimentally the male sticyleback can be induced to exhibit his peculiar reproductive patern of behaviour by introducing into his tank a dummy silhouette of a fat lady stickleback attached to a fine wire.

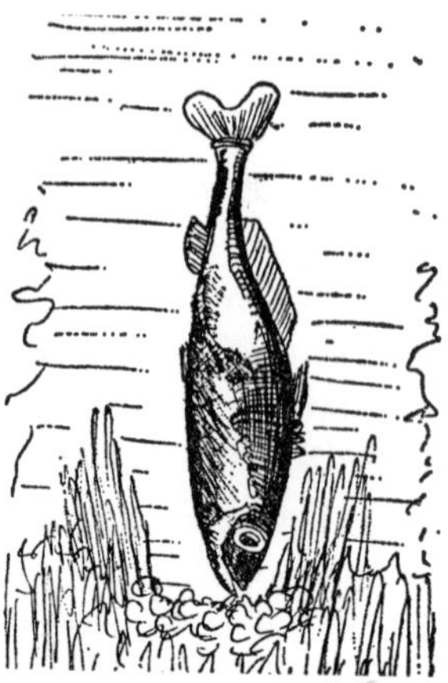

When the baby sticklebacks hatch out the male takes entire charge of them, marshals them about, protects them from dangerous currents, and in every way gives them the attention their mothers would have done had they stayed with the family.

Quite a lot of varieties of fish lay pelargic eggs, those which float on the surface and hatch in the warmth of the sun's rays.

This is a wasteful method as enormous numbers of eggs are eaten by birds and fishes, while the young ones (often carrying a black spot which makes them more conspicuous) are gobbled down by other fishes. However, enough eggs hatch to keep the species in existence and it must be realised that unless the survival ratio were low, the numbers of little fish which would grow into big fish, would very soon cause dangerous overcrowding.

The burying bettles co-operate in an endeavour to secure the future of their young. Put in this way one might gather the impression that their behaviour was purposeful, which of course it is not; it is merely instinctive and automatic. The beetles do not know what they are doing or why they do it.

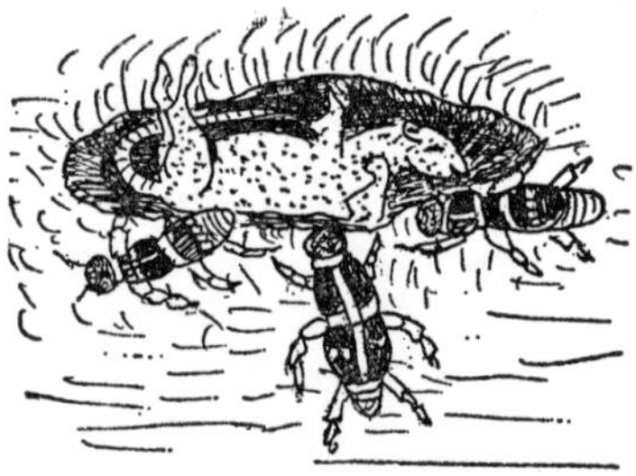

The beetles seek out a small dead animal, possibly a shrew or a rat or a mole. They burrow beneath it and scoop away the earth so that the body falls into a pit and eventually lies beneath the surface. Then the females lay their eggs beneath the body, emerge and scrape back the earth over the carcase to complete the funeral. The eggs hatch and the larvae live on the decaying body.

Nest Building

All the stages of mating, nest building, egg laying, brooding and rearing are controlled by different hormones. These are produced in the body at the suitable period and they set into operation the appropriate form of behaviour suitable to each occasion.

The ability to build a nest is inherent in each female chick and to some extent in many of the males, but it cannot be demonstrated before the nest-building hormone is secreted. This does not cause the bird to build the nest but it is the agent which enables the inherent instinct to operate so that the bird becomes possessed of the knowledge of how to weave the materials into the form and shape required, together with the desire to get busy on the task.

A good many psychologists consider that in nesting, egg laying and incubation, the completion of one stage of behaviour is the stimulus for the commencement of the next stage but such a theory makes no allowance for hormone secretion, which varies in its chemical composition from time to time. It is this change of hormone which triggers off the corresponding change in behaviour by awakening the mind of the bird to the type of instinctive action which has become necessary. All the work put into the nest by the hen, with the male collecting material, is entirely instinctive. Neither bird has the least idea why it does it or what it will do when the nest is completed.

When one stage, nest building for example, has been carried out and incubation is in progress, if by some means the nest becomes damaged, the hen bird will be incapable of rebuilding it since the nest building hormone is no longer being secreted and accordingly her mind can no longer build a nest under the influence of this hormone. Similarly when the eggs hatch and the influence of the incubation hormone wanes, the bird will lose her desire to sit on eggs and will turn her attention solely to chick minding. The situation resembles one conducted under hypnosis or under that of a drug, represented by the hormone.

The male bird does not come under the influence of this same set of hormones. He does not develop the same bare, warm brood spots on his breast that the hen does and although he will sit on the eggs occasionally while the hen goes off for food, his contribution of body heat does little towards keeping the eggs warm. He merely keeps the cold air from chilling the egg surface and he makes no attempt to turn the eggs, as the hen does many times a day when she feels the warmth of the eggs a little too great against her naked breast.

The cockbird helps in the nest building mainly by fetching material, but it is likely that some of his male hormones contain a nest building ingredient, at least in the stage of courtship, since many male birds build several imperfect cocks' nests during this stage of reproductive behaviour. These are usually rather shoddy affairs and the hen bird will examine them when she arrives and may even select one, pull it to pieces and completely rebuild it to her own satisfaction.

Husbands vary in quality in the animal world much as they do in human society. Some are companionable creatures ever ready to lend a hand with the family, and always eager to find them food, while others are completely self-centered and regard the female and her progeny as necessary evils to be borne with manly fortitude.

A great many other males having served their natural purpose move on elsewhere, refusing to accept marital responsibilities for which they have no special liking.

Others are easy-going like the male pigeons, and are content to share a wife with two or three other cockbirds; or they may be like the farmyard rooster, never satisfied with anything less than a harem, never taking any active part in raising a family but making a great commotion every time one of the hens produces an egg.

Nests may be made in a hole in a tree trunk, in a rabbit hole, in a bush or tree, on the ground and in a variety of other places. They may also follow anyone of more than a dozen patterns.

Robins and blackbirds, used to being in contact with human beings,, may choose a building site varying from the pocket of an old jacket to the exhaust pipe of a tractor. This only shows that the bird is anxious to nest and that its brain is incapable of considering possible disadvantages of a site.

Dr. D. W. Snow in *A Study of Blackbirds,* describes how pairs of blackbirds sometimes divide their families into two batches, one of which is fed exclusively by the father and the other by the hen bird. Apparently, the youngsters fed by one partner are completely ignored by the other, and if young birds beg from the wrong parent they do so half-heartedly with little hope of reward. The male bird will sometimes feed part or all of the family for as long as sixteen to twenty days, the female at this time being engaged in re-building or repairing the old nest.

The male brush turkey shows some of the inventiveness of the do-it-yourself male. His mate presents him with a sitting of eggs, then leaves him for ever. The turkey cock makes a huge nest out of rotting grass and brushwood, in the way a gardener might build a hotbed for his vegetable marrows. He leaves the eggs buried in this pile until they hatch, without expenditure of marital energy. He rears the family entirely unaided by feminine assistance.

In a case of this sort it is obvious that the bird's gonads secrete a mixed male and female hormone. During the mating period the cockbird will first secrete a pure male hormone, and later a hormone largely female in composition—in other words it will contain an excessive amount of an oestrogen.

Similarly the hen bird will start off with a female hormone, an oestrogen, but after she has laid her eggs this will become diluted with a male hormone—an androgen—and she will go forth into the world with temporarily masculine ideas. When these fade and her hormones again become oestrogenic, she will find another mate and produce a second batch of eggs.

The birds themselves are quite helpless, their urges being created by their physiological condition.

All this turkey behaviour may appear an unreasonable way of conducting matrimony, but when one sees how well it works and how readily the species can increase in numbers by such a method, one realises that Nature often knows her own business best.

In a somewhat similar fashion farmyard cockerels which are either surgically caponised, or are injected or implanted with a dose of female hormone, will not grow the male comb and wattles, nor will they acquire the male colourings. But like the brush turkey they will attend to the processes of incubation and will rear the chicks with as great care and solicitude as would be shown by any hen.

These remarks on hormones may draw attention to the fact that throughout the whole animal kingdom, including man and the higher apes behaviour, apart from being inherent and instinctive, adapted or learned, or purely intelligent as the case may be, is always conditioned by the influence of a series of hormones secreted by the endocrine glands of the body.

Animals lower than man are guided in their behaviour solely by their instincts and by hormonal conditioning of such instincts. Man, being part of an organised social community, is no longer so dependent upon instinct. He has sufficient intelligence to decide for himself whether he will follow an instinct blindly, whether he will behave exactly as his hormones make him desirous of doing—or whether he will exercise his own intelligence and will-power in order that his behaviour may conform with the general rules for decent living laid down by the state of society to which he belongs.

His inclinations and his temptations, his cravings to obey his animal instincts, vary in intensity from time to time. Normally, every man is partly male and partly female but at certain periods of the year he may be almost wholly male, while at others he may be less virile and apt to develop emotions more feminine in their nature.

Similarly, women secrete a certain amount of male hormone and their adrenal glands possibly play a greater part in their character formation and even in their appearance than is the case in the human male. If the secretion of male hormone becomes excessive the woman becomes flat-breasted, develops hair on her body and face and may sprout a beard and moustache.

Sex reversion of this sort is extremely common in birds. The farmyard hen after a useful life of egg production may begin to grow a comb and wattles, develop male plumage, commence to crow and eventually adopt a harem. It is unlikely she will prove fertile but her inclinations will be wholly male. So potent is the male hormone that chickens injected with it at two days old, develop combs, crow and attempt to mate with their own mother.

For further reference to the rearing of young, and general reproductive behaviour I would refer the reader to *The Female of The Species* (Country Life Co. 1960).

For reference to the various means by which animals communicate with one another, the reader may find much information in *How Animals Talk* (Country Life Co. 1959).

Chapter VI

ANIMALS AS INDIVIDUALS

It would be interesting if we could determine the stage of the young animal's existence at which it first experiences consciousness, and commences life as an individual.

A good many physiologists will assure us that it cannot be until after birth, since the blood supply to the brain of the foetus is not oxygenated by the young animal until after its lungs have filled with air. This happens directly it is precipitated from a warm uterus into the cold air, a fact which sets a reflex into action and causes the young animal to gasp, thus filling with air its lungs, which until now have been collapsed and empty.

But while in its foetal state within the uterus the foetus was taking into its blood sufficient oxygen for its needs by transfusion through the placenta of oxygen contained in the blood of its mother. The oxygenated blood was being pumped through the body of the foetus by the action of its own heart.

The question arises as to whether the amount of oxygen entering the blood in this way is sufficient to supply an active brain. That the brain itself is fully developed is shown by the fact that calves born at seven months gestation, two months befor their normal time, are mentally not inferior to calves born at full term.

Another way of putting the same question is to inquire whether the young animal just prior to birth has perceptivity of skin sensation or of inflicted pain.

When a veterinary surgeon makes an examination of a pregnant cow he introduces his hand and arm into her rectum and feels the calf lying within the uterus in front of, and below his hand. Although separated from the calf by the walls of the rectum and uterus, he may feel a foot of the calf and grasp it through these walls. If the calf is living it will immediately attempt to jerk the foot out of his grip. This is a reflex action stimulated by skin contact. It does not follow, of course, that a reflex need always be associated with conscious perception, in other words that the message is carried to the brain of the calf in the form of sensual perceptivity. But we also are aware that a calf anaesthetised after its birth, does not show this reflex even when the skin is cut by a scalpel. The inference therefore is that the unborn calf in the final stages of gestation can feel when its skin is handled.

In cases of difficult parturition in the cow before Caesarian section became standard practice, oversized calves, and some badly presented, were cut up inside the cow and removed piecemeal. Many of these were fully alive while the operation was in progress, but it was maintained always that calves felt nothing before they commenced to breathe after delivery.

Whether the unborn calf is capable of feeling pain has never been answered satisfactorily.

It is interesting to note in this connection that a great many people have at one time or another claimed some recollection of their late intra-uterine existence, and although such memory must be dim, since the intelligence of an unborn child cannot be highly developed, it lends further interest to the question of at what stage of gestation the foetus develops consciousness and first becomes an individual?

One thing is certain. Directly any young animal commences its extra-uterine existence it is already highly reflex-conscious, and if its behaviour at this age is entirely instinctive there are, nevertheless considerable variations met with in the behaviour of different individuals within the same litter.

I have mentioned previously that in canine circles the dominant puppy demonstrates its status from the very beginning of its existence. Although it cannot see and has to locate the teats by turning its body round in short circles, using its own posterior as their centre, it eventually stumbles upon a teat, aided greatly by its smell.

The puppy who is to be the dominant member of the litter

will work its way back into the warmth of its mother's groin and will grab hold of the last teat on the side uppermost at the time. This is the most productive teat of all and the puppy will hang on to it against all comers.

The weakest puppy will take over one of the two anterior teats and it is likely to be pushed forward by all the pups in turn, and also possibly by the mother as well, until it has been thrown out of the nest to die of starvation and exposure. It may be well that such a puppy is ill, and has not the normal smell common to the other puppies.

We will continue to discuss the progress of puppies because we are on safe ground. Most of us know something about puppies but we might become confused if we switched over to the private life of some less known animal, such as the dolphin. Although the story in either case is very similar, each mammal has its own little peculiarities to fit in with its own special environment and the circumstances of its life. For instance, the suckling of the young puppy on its kennel floor, and that of the young dolphin below water, will each require special modes of adaptation which need not concern us in this volume.

The instinct of the bitch induces her to eat some portion of the membranes which enclosed the puppy when it was delivered. This serves two purposes: it ensures that the airways of the puppy are no longer enclosed, so that it can breathe freely. It also sets into operation another urge on the part of the mother—to lick the puppy all over. The licking cleans and dries it, and also sets into motion reflexes within the puppy which cause it to pass its motion, as well as its urine.

The act of licking establishes contact between the dam and the puppy, so that the bitch accepts it henceforth as her own.

All mothers do not develop this maternal complex in regard to their own infants, and its absence has given rise to many tragedies. In such cases something has gone wrong with the normal processes of birth. The mother may have been disturbed mentally, unduly worried, without proper preparation for the travails of birth. After the delivery she may fail to recognise her infant, and it may not contrive to suckle. Suckling also helps to establish the maternal complex.

Human mothers do not consume any part of the afterbirth, nor do they lick their baby. They usually receive human assistance and may sometimes not see the child until some hours after its birth. In such cases where the contact is disturbed the mother has to rely upon her own intelligence to take the place of normal responses. Sometimes after this interval suckling does not definitely or immediately establish the feeling of ownership, especially in the case of the first baby.

When we compare unusual human behaviour with that of other animals we see a considerable degree of resemblance. A heifer, a comparatively normal female mammal, which has not yet been greatly disturbed by the interferences associated with domestication, quite frequently fails to recognise that the sixty pounds of wet struggling calf lying on the ground beside her, is in any way connected with herself. She may show as much fear of it as though it were an unrecognisable wild animal which has suddenly made its appearance. She may feel no desire to lick it, or consume any portion of the membranes surrounding it, and instead she will exhibit aggression. She will probably attack the calf and may end up by tossing it over the hedge into the adjoining field, then going away and leaving it.

Farmers are quite aware of this "heifer complex." Their remedy is to bring along the farm dog. The heifer at once becomes protective and attacks the dog in place of the calf in defence of the young animal she had not previously recognised. This is an instance when the introduction of the second response automatically engages the first normal response, the sense of maternal possession.

This protective reaction is so strong that some more elderly cows can be induced to adopt a succession of calves other than their own, by the simple method of first introducing a strange calf which is not well received by the cow, and following this up by introducing the dog. By this method one cow will often raise, or partially feed, four beef calves.

Other animals, of both herbivorous and carnivorous species, may sometimes kill, or kill and devour their young ones, particularly if any disturbance distract their attention from the immediate care usually given to the young ones directly they have been born. In addition, if the young are handled by a human owner and their normal smell is replaced or disguised by that of the human hand, this may be sufficient to cause the mother to reject or kill her young. This applies even to animals such as birds and to tame rats and rabbits, chinchillas, mink and similar animals kept by fanciers. It may affect animals as different as the cat and the sow.

A female animal at the time of parturition, so far as the acceptance of the newly arrived infant as a part of itself is concerned, is guided just as a human mother is by her senses, but in the case of the lower animal scent takes precedence over sight. Moreover, the animal at its first parturition lacks the advantages afforded to the human female of pre-natal instruction. She relies entirely upon instinct.

If a black puppy be taken away from a bitch nursing a litter of puppies, she will at first show some distress and may leave her litter for a while to search for it. If the black puppy be taken out of her hearing distance, then placed in a basket with a white puppy, the property of a different bitch, and if the two are left together for half an hour until the white puppy has acquired the odour of the black one, the bitch may accept the strange white puppy in place of the black one taken from her. She will examine it carefully with her nose, turn it over and smell its underparts, and may accept it as her own puppy without recognising the difference in its colour. Later in the day she may also take back her own black puppy, apparently unaware that her family has thus increased in number.

I am indebted to Alistair N. Worden, for some of the following observations on the two following pages, abstracted with his permission, from a paper presented by him to the Congress of the British Veterinary Association, in 1959.

The young animal demonstrates a need for something to suck, and if this does not take the form of the mother's teat, it will seek a substitute. When it sucks in this way the act will be non-nutritional. Calves, pail-fed, will suck the ears, navels and prepuces of other calves, and will even suck their own navels.

Children suck their thumbs, and are usually supplied with a dummy teat.

Levy, D. M. (1934) showed that puppies fed from bottles having small holes in the teats, sucked for an average of eighty minutes a day, with no displacement sucking. Puppies fed through nipples having holes so large that their food requirements could be met in thirteen minutes a day, exhibited post-feeding displacement sucking which continued even during sleep.

Scott and his colleagues (1958) divided the development period in puppies into four periods, varying from individual to individual within relatively narrow limits.

(a) A neo-natal period—from birth to opening of eyes, about ten to twelve days.

(b) A transition period—from opening of the eyes to time of ability to hear sound at twenty days.

(c) A period of socialization—from twentieth day to weaning at seven to ten weeks.

(d) A juvenile period—lasting from the seventh to tenth week to age of sexual maturity, six months, or later.

The period of socialization also covers the "tasting period," during which the puppy samples the flavour of everything available. Infants have a somewhat similar habit of popping everything portable into their mouths whenever this is possible.

Until the twentieth day of life—the end of the transition period—the puppy cannot develop conditioned reflexes, or learned responses. Electroencephalograms first indicate a difference in the waking and sleeping mind of the puppy at the twentieth day of life, and the adult electroencephalogram first makes its appearance at the seventh or eighth week.

Wolf puppies adopted by humans before their eyes open accept domestication, but if captured a few weeks later are difficult to tame.

Kittens born wild from domesticated parents in country districts, revert to the wild state quite readily. If they are caught it may be many weeks before they can be handled with ease. If they secure their freedom later they will usually again "go wild" during the breeding season.

Thorne (1944) showed that about seventy three per cent. of the progeny of the female decendants of a shy fear-biting bitch retained the tendency, even when fathered by a docile dog. Not only are fear and shyness hereditary but the puppies from a shy bitch learn from their mother to react as violently as she herself does.

Scott and Fuller (1958) found that Basenji puppies showed greater shyness at five weeks than cocker spaniel puppies did at the same age. He considered that this shyness was inherited on Mendelian lines.

Young dogs commence to cock the leg during urination at the sixth or seventh month, and from then on commence to take an interest in sex odours. The dog certainly starts marking out his territory quite early in life since he has an objection to soiling the territory itself, but outside it he has no great selective instinct and responds to suitable odours by cocking his leg, even when miles outside it.

Although aggressiveness is detrimental in any dog or other animal kept in domestication, it has been shown that some slight degree of aggressiveness is essential in dogs required for exhibition purposes. Character goes a long way in the showring and unless a dog is sufficiently interested in the ring and its inmates to be prepared to tackle any dog present he will usually accept the alternative, which means that he will carry his tail between his legs, his head low, and his gait will degenerate into a shuffle. The slightly aggressive dog trots into the ring full of life, eager to encounter other dogs, with ears moving and features expressive. Conformation, however good, can get a dog nowhere unless he displays it to advantage.

Jealousy and possessiveness are characteristics more commonly seen in gundogs. These are more likely than others to take possession of the best armchair and dare any member of the family to dispossess them of it.

Cats in the wild state and to a lesser extent in domestication, defend their territory against other cats, and will even attack dogs, often with considerable success.

Young kittens introduced to a new home will usually explore thoroughly every nook and cranny of the room in which they are confined.

Not infrequently various cat territories may overlap somewhat, especially in urban districts. Male cats fight fiercely when guarding territory or arguing over a female. In fighting they endeavour to grip the back of the neck, as the large cats do when killing their prey,—or they bite the root of the tail while the other participant is in full retreat.

Female cats frequently carry their kittens by the base of the neck, one at a time, to fresh quarters, and in similar fashion they retrieve kittens that have strayed from the nest. Retrieving is brought about by an innate releasing mechanism triggered off by the wailing note of a lost kitten. Bitches, lionesses, tigers, stoats, bears, and many other animals, behave in similar fashion.

I have a record of one retriever bitch which travelled two miles across a town, retrieved her puppy from a large saucepan in which it had been "drowned," carried it home, and rescussitated it.

Lorenz (1954) wrote, "I am perfectly certain that without previous experience, cats do not understand the expressive move-

ments of dogs, even although they are so similar to their own . . ."

Personally, I find this applies equally to young dogs who race up to a spitting cat without fear or suspicion, when it is standing with back arched and claws unsheathed, and appear quite unconscious of the significance of the display.

I think that only quite young kittens place any trust in a strange dog. The experienced cat, especially one that has borne kittens, usually gives the dog a chance to attack, then shins up the nearest tree, or over the garden wall. Occasionally a cat will exhibit aggression, jump on a dog's back, and attack with tooth and nail. Very few dogs can find the answer to this form of warfare.

Many other animals do not anticipate trouble at first but when they have suffered an unpleasant experience never run the risk again if they can avoid it.

During the routine testing of cattle with tuberculin, many herds a good many years ago accepted their first test, which involved intradermal injections, without manifesting undue fear. After the first one or two experiences cattle commenced to fidget, stamp and pull back on their chains or bails, immediately the veterinary surgeon entered the premises.

I have recollection of one Guernsey bull which commenced to roar every time my car entered the farm, when it was still quite a hundred yards from the bull shed, although the bull could not see the car or its owner.

Monkeys and dogs owned by me about this time, all used to cry out when I was some distance from my own gates in this car, and as we had another of the same make and pattern to which none of the animals paid any attention, I can only imagine that the car in question produced some kind of supersonic whistle which our own ears did not recognise.

Can Animals Reason?

Although it is strenuously denied by animal psychologists that any animal below the primates is capable of reasoning, all of us encounter at times behaviour in animals for which it would be difficult to find a simpler explanation.

I will quote a few instances from my own experiences of such happenings. Thousands more could be quoted by other observers if the accounts could be collected.

I was in the habit of fishing in a small lake about two miles from my home in Cornwall. On this occasion I had walked there and back accompanied by my Airedale dog, 'Mac.' As I was recovering from a damaged ankle at the time I carried a stout, cherry stick, something I do not do on other occasions.

While we were having tea on my return, with Mac apparently sleeping on the mat by the fireside, I said:

"I've just remembered something. I left my walking stick down at the lake!"

Someone present commented that I should never see it again. No other remark was made. An hour and a quarter later Mac walked into the room with eyes lit up, and tail proudly wagging.

In his mouth he bore my stick. This was no casual jaunt as to reach the lake Mac had to negotiate the traffic through a busy town both there and on the return journey.

I was losing a quantity of cooked meat nightly from my kennel kitchen, and decided to keep watch. The kitchen door was fastened by a Yale lock, of which I kept the key. The only access to the kitchen was through a gap an inch deep between the bottom of the door and a concrete floor, through which I did not think a rat could pass.

I fixed a dimmed light above the door to shine down on this crevice. Soon two adult hedgehogs appeared, and trailing behind them came two babies little larger than fat mice, still in their soft spine stage. These wriggled under the door while the adults lay on their sides waiting. Presently the adults commenced to scramble with noses and feet, and eventually withdrew a small lump of meat. This performance continued for a quarter of an hour. Every now and then one of the adults would grab a lump of meat, carry it across the yard and disappear with it beneath the floor of a range of kennels.

The following day we carried out some excavation work and found two pairs of adult hedgehogs and four baby hedgehogs together with about five pounds of meat.

The procedure was simple, but how did the baby hedgehogs learn what was required of them?

I carried the hedgehogs some miles into the country and deposited them in a suitable hedgerow. They did not return.

One of my friends, living in Cornwall, lost his mongrel terrier dog.

As there were countless disused mine shafts in the vicinity, down which many dogs disappear in pursuit of rabbits, we visited a number of these without hearing the sound of barking.

The dog had a friend, a smooth fox terrier bitch, and the two had been in the habit of going off hunting together. Someone noticed that the bitch went off alone, always travelling in one direction. This led us to an old shaft from which we could distinctly hear barking. When a rescue party descended about forty feet they found the dog on a ledge, uninjured. Also on the ledge was a fresh bone which the owner recognised as that from a leg of mutton of the previous Sunday, together with several bread crusts.

As no person had visited this shaft it can only be presumed

the bitch carried them to the mouth of the shaft and dropped them.

I have a record of a tawny owl which was discovered injured among bracken in a plantation where it had obviously lain for several days. Around it, but untouched were three dead field mice and a partly devoured frog. The bird could neither walk nor fly, so it is unlikely it caught these itself. It also seems unlikely that at the time of its collision with an overhead cable it would have been carrying such a load all at one time. The mice did not appear to have been vomited up.

Two of the local farmers assured me that is was common for owls to carry food to a sick or injured member of the species. Concerning this I have no information.

The following story does not illustrate cleverness but rather the reverse, in an otherwise intelligent animal. It shows how under unusual circumstances the mind may fail to respond.

The subject was a twenty-eight pound Rhesus male monkey which was in hospital after amputation of a necrotic portion of its tail. To keep his teeth from the sutures I had fitted around his neck a plywood Elizabethan collar, about eighteen inches in diameter, which I had great difficulty in keeping on him, as although he rather liked wearing it he could remove it with his hands without much difficulty.

While being exercised on the lead without his collar, he suddenly picked up the collar, dived into an empty dog kennel, and managed to pull the collar inside. Once inside, he slipped his head through the central aperture of the collar—and then discovered that while wearing it he could not get out again. This was a predicament, but as the kennel was roomy inside all he had to do to get out was to take off the collar. He could not be persuaded to do this, nor would he accept help.

In consequence the kennel had to be carried into a shed and locked inside it.

Thirty-six hours later the situation was unchanged, so I decided to remove the roof of the kennel. As soon as this was done the monkey hopped out, grabbed the edges of the collar and extricated his head with the greatest of ease.

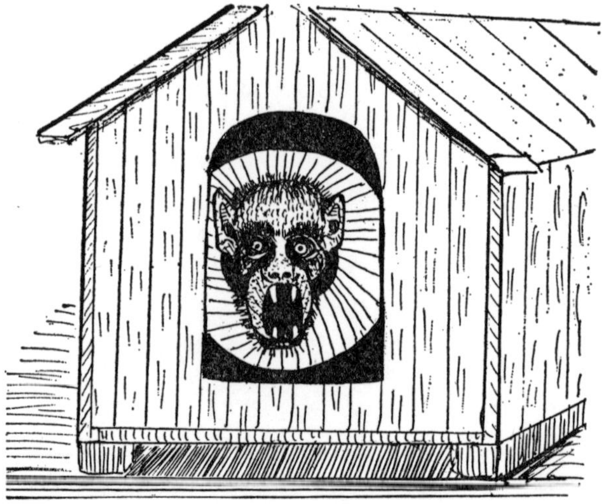

I still do not know whether the monkey's sense of humour encouraged him to prolong the joke, or whether his intelligence did not provide a perfectly simple solution.

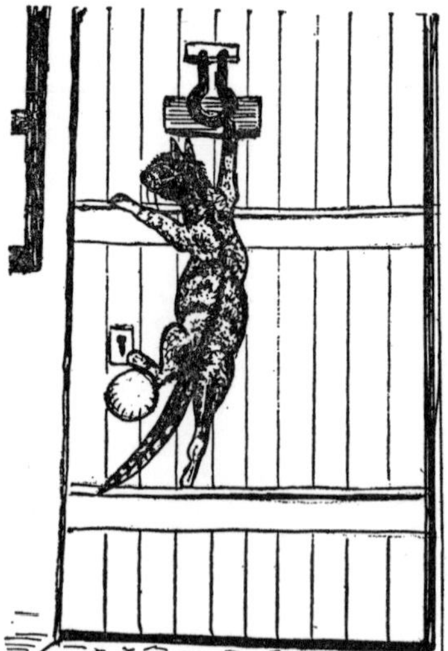

When I was a lad I owned a female tabby cat which I treasured. I was always grieved when the family insisted that Tibby must be shut out at eight o'clock every evening before I went to bed.

My elders were surprised on several occasions to hear a loud rap on the door knocker at nine o'clock, and although they did not connect the two events, they were equally surprised when nobody was in sight, but Tibby, obviously waiting, slipped indoors quickly and ran up to my bedroom.

Eventually it was discovered that after standing well back from the door Tibby would take a run and a flying jump onto the knocker, then lifting this away from the door she would drop to earth—and await an answer.

Mr. Roblin, a retired schoolmaster of Nottingham, owned a Sealyham bitch and her litter of puppies. He was in the habit of letting the bitch out of an evening, shutting the puppies in a house in his garden, which they shared with their mother. The door was closed by an ordinary flat iron bar which dropped into an iron slot on the doorpost.

My friend was surprised one evening to find the bitch out with the puppies, but concluding he had forgotten to latch the door he shut the puppies in the house and dropped the latch into its slot. An hour later the puppies were again out with the bitch. When he shut them in the house this time he saw the bitch walk up and lift the latch with her nose.

He replaced the bar latch with an eye and staple lock with a peg in the staple. The bitch found no difficulty in opening this either.

He then substituted a sliding bolt with a safety catch, so that when the bolt was in position the handle turned downwards and was held in place by an iron projection so that the handle had to be turned to an angle of ninety degrees before the bar could be pushed back.

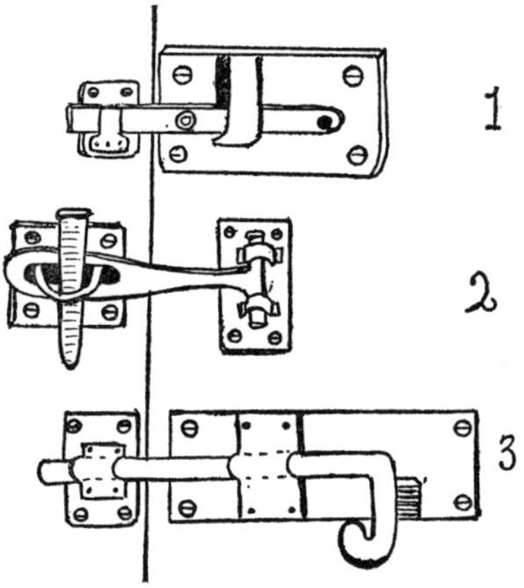

When this had been fitted the bitch sat and gazed at it for a quarter of an hour, her puppies being inside the house. She walked up to the door, lifted the handle of the bolt to the ninety degrees with her nose, then flipped back the bolt with her left front paw. And out came the puppies.

ARE ANIMALS HAPPY?

How many of us really understand what is meant when we speak of happiness as it affects animals other than ourselves? We cannot see a great way into the minds of wild creatures and our calculations have to be made largely from observation of animals which share our lives in domestication, or from laboratory animals, which are quite a different kettle of fish.

We are all well aware that the kind of life that makes one person happy may be sheer purgatory to another. It has been carefully explained, however, by students of animal behaviour that this does not apply to the lower creatures who are not able to think and reason and, accordingly, take all conditions of life as they arise, without further consideration of whether they are good, bad, or could be bettered. You are not forced to believe this unless you feel so inclined.

Is happiness the state in which a person or animal is not consciously miserable, or is it one in which the individual is exhilarated by the appreciation of circumstances which afford pleasure or anticipation of pleasure soon to come?

It may be presumed that animals, like human beings, may be passively happy when they are not actively unhappy. Similarly, it may be supposed that they may be passively unhappy or actively unhappy, merely a matter of degree, perhaps.

It would seem that the majority of people living normal lives, remain for the most part either passively happy or passively unhappy, and that they seldom find cause to become wildly exhilarated over fresh happenings, although they may become actively unhappy over the sort of events which disturb all classes of animal whether they live in a council house or in a rabbit warren.

Human beings suffer the dubious advantage, so far as happiness is concerned, of being able to think and argue with themselves. We remember things that happened recently, or quite a while ago. We are also very conscious of what is happening at the moment but we have a far more eager anticipation of things which we hope either will, or will not, come to pass in the near or distant future.

Animals have probably a greater appreciation of what is happening at the moment and realise whether the proceedings cause them pleasure or annoyance, but I do not believe that animals review the past and rejoice, or fret, over things which happened some little while ago. *We* fight our battles over and over again, think of the things we said or did, and then of the things we wish we had said or done instead. Out of it all we experience either a renewed glow of satisfaction—or we kick ourselves over and over again because we did not make the best use of the occasion.

Animals do not, I believe, rake up the past, for lacking verbal recollection they are dependent upon mind pictures which even in our own brains may be very fleeting and quickly forgotten. Neither has the animal any marked anticipation of what is likely to happen the day after tomorrow—it may anticipate the culmination of its immediate purpose, but when that is attained its mind is open for fresh adventure and it is at this stage of its activities that the termination of one piece of behaviour may set in operation the responses which set the next piece into action.

While animals are impressed, probably far more than we are, by the doings at any particular moment, it is unlikely that many species posses a time sense concerning the past or the future. Travellers who have been cast away in surroundings where they have been without clocks or watches, have commented on this fact and indicated that time in its mathematical sense gives way completely to other indications of the moment—an empty stomach or the approach of darkness.

But if a child has been brought up by brutal, dissolute parents, is used to being whipped and ill-treated, a permanent impression will be made on its mind associated with an active unhappiness tempered by a feeling of insecurity. One can observe precisely the same picture in the case of a puppy which has been similarly treated. The difference is that if the puppy is taken away from its cruel owners and introduced into a kind home, within a very few days the puppy will have become a normal, happy little creature. Its reactions to the approach of its owner will have changed from cringing to joy, and it will become obvious that its past life has been forgotten, or at least, its memory disregarded.

The child, on the other hand, will probably never exhibit this completely changed attitude to its foster-parents and the experiences it has undergone will have made a permanent impression, one which will colour the whole of the child's future behaviour.

While animals probably remain passively happy through the greater portion of their lives, they also have their moments of exhilaration, usually associated with the satisfying of an appetite, physical or sexual. No doubt young animals derive similar satisfaction from play.

It does not follow that animals are any less happy because they forget the past or that they lose a great deal through being unable to look ahead. In our lives looking back is so often tinged with regret or longing, and the ability to prognosticate accurately would undoubtedly be included among the mixed blessings.

Most of the animals we are able to study are those which share our homes, our stables or fields and spend their lives with us in a greater or lesser degree. Even among these we cannot fail to observe great variation in intelligence and temperament.

We are far less able to closely observe the behaviour of those wild animals which conceal themselves underground, or in cover, as we can usually view little other than their routine patterns of behaviour and although every dog may exhibit certain differences in its behaviour and temperament which one cannot fail to notice, to most of us one mouse behaves very much like another, and when we journey into the jungle we find that the behaviour of one lion is very much like that of another also, while that of the lion's victims seems to follow the pattern peculiar to the species without any variation; nothing, in fact, which might cause one to believe that certain individuals exhibit even slight differences that are indicative of unusual intelligence.

Variety of behaviour it would seem is introduced in the domesticated animals by the influence of their human contacts, either because the animals become conditioned to a type of behaviour acceptable to their owners, or because the environment they have entered and the kind of lives they are expected to live necessitate a particular type of ordered response.

What we would like to determine is whether animals are more happy in domestication than they are when they are living natural wild lives in the open, or less happy. A great deal, of course, depends upon the species and also upon the animal's individuality. Some animals captured in the wild, even when adult, become tame very quickly, and appear to enjoy life under conditions of domestication, to such an extent that when offered their freedom they prefer to return to what we term captivity.

A few weeks ago I came across a woman at the door of a country cottage who appeared to be having a fierce fight with an aged rook. She was taking wild swipes at the rook with a broom, while the rook was dive-bombing the woman as she stood there filling the cottage doorway.

When I went to her rescue I found that my efforts were being devoted to a very unworthy object.

After we had finally closed the door, leaving the rook without, I learned that the woman had found the bird in the roadway with a damaged wing a month earlier and had taken it in and fed it and looked after it until it was well again. Then, when she had attempted to give the bird its freedom it flatly refused to leave the house.

After this explanation I took my departure, and as I walked out the rook flew in.

There is of course a great deal of difference in what we term captivity—open captivity, in which the animal is confined voluntarily or involuntarily to the territory administered by its protector— and the kind of captivity of the closed zoo where the animal is confined in a cage and is incapable of being handled and talked to. The human voice for some curious reason has a great fascination for many kinds of animals.

Generally speaking it is not easy to tame animals which have already lived the greater part of their alloted span in freedom, especially when they have never previously regarded the human being as anything other than a potential murderer. It requires a great deal of patience and particular state of mind before one can obtain the confidence of such creatures, although in a good many instances one may succeed. One may then discover that the animal shows a great deal of affection towards its new owner and exhibits every sign of being completely happy in its strange surroundings. How much of this is due to the animal's own sagacity and to a recognition of the fact that its own economic position is now assured and that human beings have a very varied and gratifying taste in foodstuffs, is still uncertain.

I am, however, inclined to give the animal the benefit of the doubt and to believe that every animal values both physical contact and affection, and that a great many, if they were in a position to choose, would put the affection before appetite. Once again I am laying myself open to all the eager hunters of the anthropomorphist.

One may argue both ways. It may be that the animal who is willing to exchange its natural life for one of semi-captivity in human company, with a satisfactory supply of food guaranteed, is showing intelligence—or it may equally well be argued that such an animal is lacking individuality, is downright lazy and a disgrace to its kind.

The ability to accept domestication, as a way of life, appears to vary a great deal, not only in certain species but also among individuals of that species. But, during a long life among animals I have encountered people who have made pets of almost every kind of animal, including snakes, crocodiles, spiders, eagles and owls, and I have even met a lady who kept and regularly fed a huge yellow slug in her kitchen. These are only a few examples but in each of these there has been a similar apparent exchange of affection, and complete understanding between the pet and its protector.

Even when taken from their nests or from their mothers, and reared by hand, one occasionally comes across an individual which stoutly refuses to submit to domestication—one in which the call of the wild is so strongly evidenced that in no circumstances will it form a friendship with any animal other than one of its own kind.

In connection with young animals taken from the nest, or from the mother immediately after birth and raised by human beings, Lorenz made some very interesting observations on what he called "imprinting," in the case of young birds. He found that a gosling, for example, which first sets eyes after hatching upon a human being, will adopt the person as its parent and for ever after will imagine that it is itself a member of the human species, to such an extent that when it attains maturity it will show a desire to mate with the person who has raised it. Lorenz found that all kinds of birds do not adopt this imprinting behaviour. Snipe and plover, for instance, behave quite differently and endeavour to escape immediately.

Portmann stated that the more embryonic any animal was at the time of its birth the more plastic it was so far as adoption was concerned. To prove this David Katz instanced the case of a child adopted by a she-wolf. This is quite a possibility and more believable when we learn more about the wolf's maternal instincts, and it is no more peculiar than a case under observation at the present time where a cocker spaniel has dug a litter of baby rabbits out of a hedge and nursed them as her own; nor of the monkey, the story of which I have recorded elsewhere, who shared the upbringing of a litter of cocker spaniels with my little cocker bitch, and insisted in doing her share of the nursing, one puppy at a time, on the roof of my dwelling house, using the breast of the chimney as a convenient backrest.

It would be interesting if we could assess the degree of happiness experienced by any of these foster babies including the human one, and determine whether they were missing any of the joys of life, or if they were completely contented with their change of circumstances.

But, the majority of wild animals converted to domesticity

when young, will show a tendency when they attain puberty, or adult age, to seek a mate. They are then extremely liable to wander afield and this applies to animals such as foxes, badgers, squirrels and the like.

I reared a dog fox from the day when I found it at only a few weeks of age being multilated by a boy with a short sharpened stick, and it became quite as affectionate as any dog. I estimated that Bonzo travelled nearly thirty thousand miles on the passenger seat of my car completely devoid of any form of restraint. He occupied the best armchairs in the house and would only give the one he occupied over to a stranger on condition that he sat upon my knee.

And yet when he became adult he left me to go on courting expeditions on the cliffs eight miles distant, coming home only at intervals to be fed. It was on one of these trips through busy motor traffic that he was run over by a car and killed.

The vixen cub I obtained in his place would have nothing to do with me, and showed quite clearly that domesticity was not her line and that all she wanted was the wide open spaces.

Imprinting appears to affect many young ungulates. Young lambs and baby deer which have been fostered from birth by a human being often show a great desire to be regarded as real persons.

There are innumerable instances, of course, where a person, (and this includes the writer) has reared a monkey or a chimpanzee on terms of mutual affection until it has attained the age of puberty, somewhere between four and five years of age. After that it has suddenly turned upon its owner and inflicted serious injury.

All the evidence points to the belief that the majority of animals which cheerfully accept domestication remain completely happy for the greater part of the time. They get food without the trouble of seeking it, and they are free of their natural enemies, and so can enjoy security and almost complete laziness. But experience also shows that happiness in animals, as in humans, demands a normal sex life, and that frustration can set into operation all the mental conflicts and unaccountable forms of behaviour which enliven or blight the lives of more intelligent animals.

It must be conceded that in the lower animals these mental upheavals occur only during the seasons devoted to lovemaking. It has been said that man is the only animal that drinks when he is not thirsty and makes love all through the year.

Looking at the matter of happiness from another angle it is pretty obvious that whether a man, a dog or a kangaroo is likely to get the most happiness out of a humdrum existence must depend in a great measure upon the individual's temperament.

From among the animals which live natural lives in the wild it is difficult to draw any sound conclusions. We cannot really decide whether some badgers are more grumpy than others, whether lionesses are given to nagging, or whether black mambas are regarded as rays of sunshine in the snake world.

Before we can gain any real knowledge of what goes on in the family circle one must share it.

But, we are in an admirable position to obtain first-hand knowledge of temperament in our domesticated animals, our dogs and cats, horses and farmyard animals, as well as among the pets we adopt.

One thing we do learn from them is that animals are no more reliable temperamentally than human beings, and that in the matter of cussedness and general unreasonableness there is little to choose between them.

In dogs, temperament as we know it, is closely linked up with domestication and in our modern breeds (three hundred of them) the characters they exhibit have been determined by selective breeding and are not the outcome of promiscuity.

The word "temperament" in their case describes the kind of responses dogs make to the human beings with whom they are associated, as well as to other animals of their own kind. It also applies especially to the manner in which dogs respond to human control and to a certain amount of discipline. From the dog's own angle it must be accepted that the temperament and behaviour of the persons in control may also leave something to be desired. The dog, being naturally a highly sensitive and discerning quadruped may easily be influenced temperamentally and made actively unhappy by harsh treatment or injustice, and cases of induced neurosis are quite as common among the modern dogs as among the elite of Mayfair. According to the advertisements in the journals which deal with canine matters it would seem that quite a number of people are setting up in practice as exponents of canine psychiatry.

The dog began as a member of a pack and his love of company has remained an essential part of his make-up. He accepts one member of the human household as the pack-boss and while he is loyal to the pack as a whole his own status is on a par with that of all the other members of the household.

The dog who is anybody's plaything is the one who has eaten too many free meals and has degenerated socially and mentally.

But dog, like man, is ruled largely by his hormones, and

individual characteristics may well keep pace with the output at the moment. Many of these have a sexual basis and this may account for the bully, the street walker and the over-ferocious watchdog.

It is probably on account of the intermittent and sometimes unexpected output of hormones that temperaments in animals are apt to prove unstable and that many animals from cats to elephants, are apt to change their relationship towards their human associates without warning.

In this connection I must mention "Jonathan," a Friesian bull owned by one of my farmer friends.

I think bulls are by far the most temperamental of all our common domesticated animals and in their case there is little doubt that the manners they will exhibit on any particular day will depend upon their current output of certain hormones.

The docility of Jonathan was a matter for village wonderment in a district where bulls were regarded as bulls. It was truly said that the children of the village were in the habit of going into the fields where this bull was running with the cows, cuddling his legs, and catching hold of the ring in his nose they would lead him wherever they wished. It was reported that some of the younger children were regularly lifted onto his back for rides.

I had attended Jonathan on a good many occasions for a hind limb lameness and each time I had gone alone into the yard in which he was temporarily confined, to give him personal treatment. But the last visit I made to him opened my eyes somewhat.

I opened the large double gates which led into his courtyard and prepared to enter. Jonathan was resting peacefully, leaning against the further wall of the yard and appeared quite undisturbed by my entrance.

And then, entirely without warning, he dropped onto one knee. He dived his horns into his bedding and threw it into piles over his back.

By now I knew what was coming; I had seen such antics in

other bulls. With a loud bellow he was on his feet and charging down upon me at full gallop.

Fortunately for me the gate, as he emerged, bashed back upon me and imprisoned me between the door and the brick wall of the yard. Jonathan could not see me but he knew quite well where I was concealed but he had not sufficient intelligence to close the door again. Instead he went to its other side and made occasional attacks upon it. I had a few inches of room but hardly enough to make allowance for my ribs which received quite a battering. There I stayed until the workmen returned twenty minutes later.

There was a sequel even more unusual.

Three months following this incident an auction sale of stock was being conducted on the farm and the site chosen was an orchard full of young apple trees.

Jonathan, described by the auctioneer as the quietest bull in the country and the pet of the village children, was led into the ring for sale by a lad who held him by nothing but a rope halter. All went well for the first five minutes while the bidding mounted and Jonathan looked on with complete unconcern. Then, suddenly, he let out one long roar and charged full tilt at the auctioneer's rostrum.

In the next few seconds every apple tree in that orchard was laden with astonished farmers from the audience.

And so the scene remained for the next quarter of an hour when the local butcher arrived in his Jeep with a shotgun at the ready. That was the end of Jonathan, the children's pet.

It is believed by a great many people that wild animals live in a state of constant fear and that they never enjoy a feeling of security which enables them to rest in peace.

Nothing could be farther from the truth.

It is perfectly reasonable to believe that wild animals live in a state of constant awareness which is a very different thing, and it is also quite likely that they are not so conscious of this awareness as one might imagine since wild animals are far more automatic in their actions than we are. Not only is their hearing much keener than ours but they also possess that peculiar "ground vibration sense", which we upright creatures lack. Even a snake, credited with being deaf, feels ground vibrations through its chin, and is just as aware of the approach of feet as a deer whose ears turn in all directions to locate sound. As I have stated, from ground vibration animals learn not only the whereabouts of their enemies but in a great many instances they can determine from these vibrations their intentions also.

That is why we see so few animals during a country walk—our whereabouts have been determined and the direction of our approach ascertained partly from our footsteps which have conveyed messages through the ground, partly by the animal's sense of smell, and it may also be because the sound of human voices carries for a great distance when the wind is behind the approaching party.

Armed with defences such as these, and with rapid responses, animals retain their confidence, and even if life entails a great deal of moving from place to place in order to maintain survival this is not counted as a hardship since their physical ability is established on such a high plane that we ourselves can have no appreciation of the joy which normal animals must experience from their intense bodily fitness.

Normal healthy animals never go out of their way, as so many human beings do, to make themselves miserable. There are no hypochondriacs in the ranks of the wild stock and all are extroverts, concerned with the things they encounter during their wanderings, and not the least interested in what goes on inside their minds or bodies.

They also have no knowledge that their ultimate fate is death; nor are they being constantly adjured to make preparation for the event. The fox hunted by the hounds, puts all its reflexes to the best advantage to escape them, but apart from the fact that it may know from experience that another animal may inflict pain it can have no direct knowledge that if it is caught it will be torn into pieces and killed.

For a great many years philosophers have wondered (but seldom recorded their findings) whether it is better to be a disgruntled human being or a contented cow.

The majority of people are liable to display sentimentality and many of them wish to believe that animals resemble ourselves in our ability to look a short way ahead and make some plans for at least our immediate future.

Today there are indications that the balance of opinion concerning the question of whether animals are capable of conscious thought is tipping back again and that more and more thinking people are beginning to disbelieve that animals are mere automatons, that their behaviour bears no similarity to our own and that their actions are purely automatic. They are beginning to doubt a little that animals are entirely incapable of thought or of elementary reasoning and that their behaviour is organised in such a way that the completion of one act is the stimulus which sets into operation the commencement of the next.

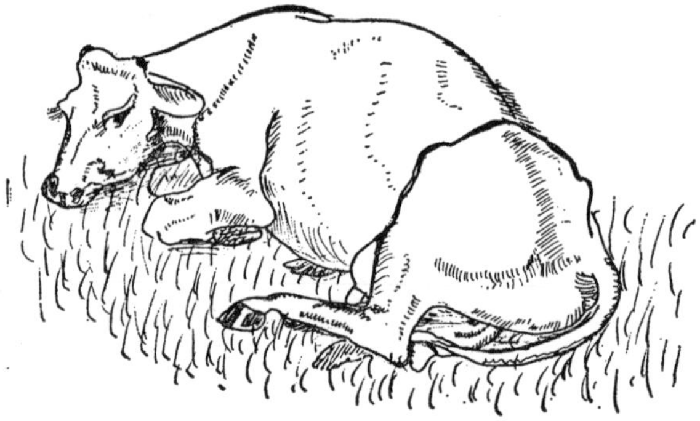

It must be admitted, however, that our contented cow can in reality have very little understanding of what goes on in her life. It may be equally correct to suggest that she is equally ignorant concerning events that have already happened during it, and far more so regarding what may happen to her in the future. Being so contented, she may be entirely lacking in curiosity in respect of either set of circumstances; she lives entirely for the moment.

There is a great advantage when an animal has an ensured food supply, a roof and shelter. Modern techniques ensure humanity a constant supply of foodstuffs in and out of season, and foods which are fresh only in summer are available in cans whether it be January or July.

Our nearest relations, the great apes, who are still without adequate winter shelter and lack cans and openers, still show an inclination to mate only at the time of the year which will enable their progeny to make their appearance during the period of plenty which coincides with the days of warmth. So far nobody has decided if their breeding habits would fall into line with our own if they, too, could be provided with houses, central heating and greater reliance upon the tin opener.

Our cow remains contented, probably because she lives on an "all found" basis and has no occasion to bother about the weather and food supplies. No such consideration enters the mind of the cow, she lies and ruminates at every opportunity. She ruminates not as a wife does, wondering what to get for dinner today, or whether the meal she prepared last Tuesday that her husband and the children appeared to like so much, might be worth repeating. She merely belches up successive mouthfuls of food from her rumen to be chewed all over again, and swallows the mass so that it may pass further along the digestive tract and be converted into a little energy and a lot more milk. She enjoys placidity and contentment, because as a result of domesticity and human control which has brought about selective breeding, her physical activity has been minimised and her lactiferous attainment glamourised, until she has developed into the perfect milking machine.

Scientific breeders discovered that the more the cow ate the more milk she secreted and that the way to increase output was to produce the cow which could eat the greatest quantity of cheap fodder and convert it into more and more costly milk.

Cows, like human beings, cannot eat continuously without resulting visible change in the bodily outline. The modern cow now displays a rotundity of figure equal to that possessed by human females in the days when they were portrayed by old masters, long before slimming came into fashion.

When any animal is provided with a lack of imagination, a really healthy appetite and a figure to correspond, life becomes a matter of eating and sleeping, both enjoyable occupations. To a lazy cow in a lazy world continuous enjoyment spells contentment.

Human beings differ from cows in their digestive accomplishments as much as in the way they regard the process of living and life's problems. People have long memories as well as certain powers of prognostication and the more intelligent among them can often exert some small influence upon the shaping of future events. They are also a little inclined to foster the belief so aptly expressed in song in that very successful musical play *My Fair Lady,* to the effect that with "a little bit of luck" many of the things they hope for may come to pass.

In spite of this no human being is ever really contented however much luck may have favoured his pursuits. He wants always just a little more, and it is this very discontentment which spurs him on to further progress. An improvement in his resources may make for a better chance of survival or it may increase the amount of pleasure he derives from the process of living.

The cow has already reached her promised land. She has attained her Cornucopia – the horn is full and flowing over. What is there left but to eat and sleep?

There are men, and women, too, for that matter, who would be quite contented if they were enabled to follow the same programme of eating and sleeping. They might introduce other diversions, but our cow wants none of them.

Sex, and the demands of sex, play no practical part in her life any longer – her reproductive needs are fully catered for by artificial insemination and she still remains contented. One may ask whether these remarks apply to the cow alone – have we still the contented bull?

The answer is that we have no bulls nowadays other than those required by the A.I. centres and by breeders who produce cattle for exhibition. All surplus bull calves are turned into veal.

We still have no answer to our question, "Is it better to be a disgruntled or disillusioned human being or a contented cow?" "Disillusioned" may not be the appropriate word for we number optimists among our ranks as well as pessimists and few of us suffer disillusionment for very long. When we suffer reverses not all of us go under when the first waves of disappointment roll up. Many have sufficient intelligence and mental energy to fight against their force and make the best attempt they can to turn them to good account. The waves of misfortune have carried many a striving spirit onto the shores of success.

Man is striving ever onwards to a visionary Utopia. The cow is contented with a negative existence in which passivity is the ruling feature. In a hundred years hence man will have altered a great deal, not in the physical sense but with regard to his mental attainment, his outlook upon life and in his relationship with his fellow men.

Providing that within the next century man has not discovered something more palatable, nutritious and economical than milk, the cow will remain as unambitious and contented as she is today. Her stomachs may be a little more capacious, she may be able to pack away a little more fodder and in return she may yield a little more milk.

The answer to the question we propounded is that contentment depends upon the mentality of the individual. Man, with his own peculiar type of brain, will never be content unless he is prepared to stagnate mentally and physically.

The cow could never remain contented if she were required to be anything other than a cow.

BIBLIOGRAPHY

Carson, Rachel: *The Sea Around Us.*

Casteret, N.: Observations sur une colonie de chauves - souris migratices. (1938) *Mammalia,* p.p. 29-34.

Harris, Sir Arthur, Bart: *Country Life.* Oct. 6th, 1960.

Katz, David: *Animals and Men.* (A Pelican Book)

Pincher, Chapman: *A Study of Fishes.*

Pumphrey, Prof. R. J.: Private Communciation.

Rothamsted Exp. Stn. Rept. 1951.

Scott and Fuller, (1958): See Worden, A. N., N.V.A. Congress Rept.

Smith, Dr. Stuart: *How To Study Birds.* (Collins)

Smythe, R. H., *How Animals Talk.* (1959) Country Life Limited.

Smythe, R. H.: *The Female of the Species.* (1960) Country Life Limited.

Smythe, R. H.: *The Mind of the Dog.* (1958) Country Life Limited.

Smythe, R. H.: *Animal Vision.* (1951) Herbert Jenkins Ltd.

Smythe, R. H.: *Veterinary Ophthalmology* (2nd Edition 1958) Mess. Bailliere, Tindall and Cox.

Snow, D. W.: *A Study of Blackbirds.* (1958) George Allen and Unwin.

Thorne (1944): See Worden, A. N., N.V.A. Congress Rept.

Vexkull: *The World the Dog Lives In.*

Worden, Alastair N.: (1959) Abnormal Behavior in Dogs and Cats. Proceedings of N.V.A. Congress.

INDEX